Final Countdown

UNIVERSITY PRESS OF FLORIDA

Florida A&M University, Tallahassee
Florida Atlantic University, Boca Raton
Florida Gulf Coast University, Ft. Myers
Florida International University, Miami
Florida State University, Tallahassee
New College of Florida, Sarasota
University of Central Florida, Orlando
University of Florida, Gainesville
University of North Florida, Jacksonville
University of South Florida, Tampa
University of West Florida, Pensacola

Pat Duggins

University Press of Florida
Gainesville/Tallahassee/Tampa/Boca Raton
Pensacola/Orlando/Miami/Jacksonville/Ft. Myers/Sarasota

Final Countdown

NASA and the End of
the Space Shuttle
Program

Library of Congress Cataloging-in-Publication Data

Duggins, Pat.
Final countdown : NASA and the end of
the Space Shuttle Program / Pat Duggins.
p. cm.
Includes bibliographical references and index.
ISBN 978-0-8130-3146-0 (cloth)
ISBN 978-0-8130-3384-6 (pbk.)
1. Space Shuttle Program (U.S.) 2. Space shuttles—United States—
History. 3. Astronautics—United States. I. Title. II. Title: NASA and
the end of the Space Shuttle Program.
TL795.5.D83   2007
629.45'4—dc22   2007013474

The University Press of Florida is the scholarly publishing agency
for the State University System of Florida, comprising Florida A&M
University, Florida Atlantic University, Florida Gulf Coast University,
Florida International University, Florida State University, New College
of Florida, University of Central Florida, University of Florida, University of North Florida, University of South Florida, and University of
West Florida.

University Press of Florida
15 Northwest 15th Street
Gainesville, FL 32611-2079
http://www.upf.com

Title page photo: Courtesy of NASA

To Lucia,
For all those late nights, long weekends, and NASA acronyms

Courtesy of NASA

Contents

Acknowledgments

Anytime an author sits down to write an acknowledgments page, there's the obvious risk of leaving someone out. So many people have lent their kind assistance during the writing of Final Countdown that the following list should be considered "including, but not limited to . . ."

Gayle Frere, James Hartsfield, and Kyle Herring of the Johnson Space Center

Dave Drachlis and June Malone of the Marshall Space Center

Bruce Buckingham, George Diller, Kay Grinter, Lisa Malone, Bill Johnson, Laurel Lichtenberger, Margaret Persinger, Jessica Rye, Manny Virata, and the whole public affairs crew at the Kennedy Space Center

And Anne Gudenkauf, David Malakoff, Greg Peppers, and Ellen Weiss of National Public Radio; and Mike Crane, Jose Fajardo, Brian Johnson, and Stephen McKenney Steck of WMFE for their support as I cover this unique, and ongoing, piece of American history.

Prologue

"Do you drink Tang?" National Public Radio's Madeleine Brand once asked me.

I get that question occasionally.

Anyone who grew up in the 1960s might remember Tang, which was the orange-flavored powdered drink mix the astronauts supposedly took with them to the moon as the United States battled the Soviet Union to put the first man on the lunar surface. As NPR's resident "space expert," I contribute about an hour's worth of stories each time a shuttle blasts off. That's in addition to whatever unmanned spacecraft NASA launches from the Cape, or missions the Russians, Europeans, Japanese, and now the Chinese undertake. That often leaves colleagues and listeners with the impression that I eat my meals through a plastic tube, NASA style.

That's not precisely the case. But in the late 1960s, I admit, I was one of those kids waiting for Tang to make its way onto store shelves. My folks were under strict instructions to keep an eye out for the jar that had the small plastic Apollo lunar rover, or moon buggy, shrink-wrapped onto it. No buggy, no sale.

Project Apollo was the stuff of pop culture back then as well as a driving force in grocery stores. Local moviegoers lined up to see *2001: A Space Odyssey*, and actress Barbara Eden incurred the ire of network television censors by baring her belly button on *I Dream of Jeannie*. Her costar, Larry Hagman, played an astronaut, and the show was set in Cocoa Beach near the launchpads of the Kennedy Space Center. The city commemorated the connection by renaming one of its roads I Dream of Jeannie Lane.

My father was an air force chief master sergeant, and my family was stationed in Alaska when astronauts Neil Armstrong and Buzz Aldrin landed on the moon during the historic mission of *Apollo 11* in 1969. The two men opened the hatch on the lunar module called Eagle around 10 p.m. Eastern Daylight Time. That meant most schoolkids had to stay up late to watch the first man walk on the moon. At our home in Anchorage in the Alaskan time zone it was four hours earlier, and we were just sitting down to dinner when the phrase "One small step for a man, one giant leap for mankind" was uttered for the first time.

In 1971 my father was transferred to Patrick Air Force Base, which sits just south of Cape Canaveral in Florida. That meant the exciting notion of being in the middle of the space program that put Tang on store shelves and Barbara Eden's navel into the nation's consciousness. I remember walking into my backyard to see *Apollo 14* blast off with Mercury pioneer Alan Shepard as a member of the crew. Unfortunately, Neil Armstrong had already beaten the Russians to the moon two years previously, and budget cuts were already chipping away at NASA. Instead of wading into a vibrant space program, I was a witness as time was running out for Apollo.

There would be just three more lunar landings. After that, the U.S. space program would surrender the moon. Astronauts who dreamed of kicking up the gray lunar soil would be reduced to fighting for a seat on the Skylab space station. The only alternative was one of the three crew positions on the last Apollo capsule, which would dock with a Russian-built Soyuz spaceship. Long after that came the troubled space shuttle. For me, it was like discovering the Beatles just as the *Abbey Road* album was coming out and Paul was thinking about Wings.

My first visit to the Kennedy Space Center didn't occur until 1976, after Apollo was long gone. Don't be surprised. Just like people who live near the ocean but never visit the beach, my family never went to the White House or the Smithsonian while we were stationed in Washington. As new Floridians, we also neglected to check out the launchpads until after NASA retired its moon program, which was simply par for the course. The agency was looking for something to do with its spaceport. Since there were no rockets going off, they used it as the stage for a science and technology exposition called "Third Century America," put together to mark the bicentennial. This was supposed to be a forward-looking affair, but with the empty Vehicle Assembly Building in the background, it was more like a monument to past glory. Something had clearly gone wrong at NASA.

A few years later the space shuttles started flying, and I was soon out of college starting my career in journalism. My experience reporting on the shuttle began with the *Challenger* disaster in 1986. Seven astronauts were killed when a leaky O-ring gasket on one of the solid rocket boosters allowed white hot exhaust to stream out and hit the spacecraft's external fuel tank. Along with reporting for my own station, WMFE in Orlando, I was space correspondent for the USA radio network out of Dallas.

NPR enlisted me to be their "space reporter" in 1996, shortly after both the tenth anniversary of the *Challenger* accident and my promotion to WMFE news director. The NPR task included providing stories and interviews for national broadcast about each space shuttle mission that blasted off from the Kennedy Space Center, as well as reports throughout the flight and for each landing.

The days of uncertainty at the Kennedy Space Center following the demise of Apollo are a painful part of NASA's history. Now the space shuttle fleet is about to be retired to clear the way for a new ship, the Crew Exploration Vehicle. According to the White House timeline, the year 2010 will represent the "final countdown" for the shuttles, and the questions have already begun on how to get back to the moon and on to Mars, which is NASA's new goal. The difference now is that the U.S. space agency seems to know where it's going, as opposed to the mostly aimless wanderings of the space shuttle.

Apollo clearly worked, because it was a mission in search of a spacecraft. The shuttle didn't work, because it was a spacecraft in search of a mission. The following pages examine this central problem in the history of the shuttle program as it looks ahead to NASA's new program, Project Constellation, featuring the Crew Exploration Vehicle, now known as the Orion capsule. It's a story of lost dreams, facing change, and new dreams to come.

Introduction to the Paperback Edition

"Houston, wheel stop." That traditional phrase among space shuttle commanders raises a lot of lumps in a lot of throats at NASA these days. It will be the last thing said at the end of NASA's last space shuttle mission after the vehicle rolls to a halt on the runway. For the men and women in the U.S. space program, that end is coming soon.

The shuttle's demise was prompted by the Columbia accident in 2003. A damaged heat protection tile allowed the vehicle to incinerate during reentry into Earth's atmosphere. The disaster sent fiery debris streaming through the morning sky and raining down on East Texas. Search teams found the wreckage, including whole control panels with the switches flipped as the astronauts had left them. The seven crew members were killed, including the first Israeli to go into space. As the nation grieved, the White House decided NASA needed a new course.

That plan meant grounding the shuttles for good, and shipping them off to museums. In place of Atlantis, Endeavour, and Discovery, the Bush administration supported a return to Apollo and the era of gumdrop-shaped capsules. A replacement vehicle, dubbed Orion, would be propelled to orbit by a larger version of the solid rocket booster that's bolted to either side of the space shuttle's external tank. The new rocket is called the Ares-1.

The plan isn't without its critics.

The Ares is under fire from space proponents who want to use tried and true rocket technology, instead of reinventing the wheel with a new booster. There are rockets that currently send people to space and there are others that launch satellites. An alternate plan is being pushed to take rockets normally used for unmanned payloads, install extra safety systems, and perch Orion on top for the trip to space. The prime candidates for this variation on a theme are the Delta 4 and the Atlas 5 boosters. Assuming NASA abandons the Ares-1, launching the Orion capsule on a Delta or Atlas could cost a billion dollars per trip. That's leaving critics in Washington wondering where to find the extra money while NASA tries to pay for its other mandates from the White House, namely, returning people to the moon and then looking at a base on Mars.

Further complicating the issue is yet another plan being offered by engineers at the Marshall Space Center in Alabama. It's a completely new rocket, also based on the current space shuttle design. That vehicle is called Jupiter and it would feature an elongated "stretch limo" version of the shuttle's butterscotch-colored external fuel tank with solid rocket boosters on either side. The Orion would sit on top. NASA is publicly dismissing all spacecraft designs that would compete with the Orion and Ares-1.

The plan to retire the shuttle fleet also turned into a political football on the 2008 presidential campaign trail. Democrat Barack Obama and Republican John McCain both made general statements in favor of the U.S. space program, but neither offered specifics in the months before the election. The issue was a high priority among Kennedy Space Center workers, who faced massive layoffs following the end of the shuttle program. Kennedy was marked early on for the most severe job cuts, as was the Michoud plant, east of New Orleans on the Gulf of Mexico. That facility currently builds the external fuel tanks that are launched from Florida. The state's U.S. Senators, Bill Nelson and Mel Martinez, lobbied both of the candidates to deliver some kind of public address on the subject of NASA.

On August 2, 2008, Democrat Barack Obama spoke before a rally at Brevard Community College near the Kennedy Space Center. The candidate pledged to protect NASA's budget and the thousands of local jobs that hinged on the space program. "Under my watch," Obama stated, "NASA will inspire the world, make America stronger, and help grow the economy here in Florida." Specifically, Obama favored extending the shuttle program by one mission and reinstituting an aerospace advisory committee to decide how to tweak NASA's future course.

On August 18, 2008, John McCain held a closed door meeting at the same community college that Obama visited. The Republican Presidential hopeful spoke with twenty local aerospace leaders, then made a short statement to the press on his support for the U.S. space program. The main issue with McCain was NASA's plan, following the end of the shuttle program, to send astronauts to the International Space Station on Russian-built Soyuz rockets until the new Orion could be built. "We need to look at all options to maintain American access to space," said McCain, "to minimize the time between the termination of the space shuttle program and the flight of the replacement vehicle."

The notion of NASA relying on the Russian space agency for access to the space station, funded largely by the United States, had already rankled members of Congress. McCain also cosigned a letter to the Bush White House asking that the shuttle program be extended until the new Orion capsule could be ready.

NASA's fate may become clearer about a year or two after the Presidential inauguration. That's when the first budgets from the new administration are expected to offer the Commander-in-Chief the chance to make concrete decisions.

While policymakers wrangle over the fate of the space program, the impending loss of the shuttle has also started a new version of the "astronaut shuffle" at NASA.

During the days of Apollo in the 1960s, astronaut Jack Lousma was assigned to explore Tycho crater on the moon. His flight, called Apollo 20, was the first of three lunar landings to be canceled due to a tightening budget in Washington.

Astronaut Gerald Carr faced a similar disappointment when NASA cut Apollo 19. That was his mission to the Hadley Rille region of the moon. Both Lousma and Carr would find flight assignments, not to the lunar surface, but aboard America's first space station, Skylab. After those trips, each man received a traditional gold lapel pin from NASA, signifying that they were veteran astronauts. But each had missed out on an even rarer memento. Apollo moonwalkers were traditionally presented with the embroidered NASA patch that had been stitched onto their lunar spacesuits. Each patch is lightly covered in moon dust.

Like Lousma and Carr, astronaut Ron Garan faced similar compromises as the space shuttle program wound down. He was selected by NASA to train as a pilot who would help the commander during blast-off and landing. The dwindling number of shuttle flights on NASA's schedule left him with a big question to consider. "We had a backlog of pilots who hadn't been assigned (to a flight) yet," said Garan, "so the word went out for any pilots who wanted to train as an M.S. (mission specialist). And I was one of the ones who got picked." Mission specialists do spacewalks, while other astronauts pilot the shuttle. During Garan's trip to orbit and back, someone else would have their hand on the control stick. "There are many ways to contribute to a mission," he quipped. Garan performed three spacewalks to help install the school bus–sized KIBO science lab built by Japan for the International Space Station.

The events that are shaping Garan's career and the future course of NASA began with the agency's Space Exploration Conference in Orlando in 2005. That's where the new Orion vehicle took its first steps toward reality.

That's chapter one of Final Countdown.

1

The Future

The laptop wouldn't work.

A room full of aerospace industry executives, members of the media, and a few former astronauts had gathered at NASA's 1st Space Exploration Conference in Orlando, Florida in early 2005 to hear a presentation on what it would take to set up a manned base on the surface of the Moon as a prelude to sending people to the Planet Mars. The problem was that the laptop controlling the PowerPoint presentation wouldn't operate and the situation was dragging on for an uncomfortably long time. Volunteers from the crowd stepped up onto the stage to help tinker with the technology, while many in the audience filled the time punching the buttons of their Blackberries. The laptop eventually did cooperate, but it underscored the message of the panel discussion led by astronaut Carl Walz, which was that the United States doesn't currently have the technology to go Mars.[1] In fact, the nation hasn't put a man on the Moon since 1972.

The frank assessment probably wasn't what President George W. Bush was looking for to support his vision of putting humans back on the moon with an eye toward an expedition to Mars. Specifically, Project Constellation calls for NASA to mothball its remaining fleet of three space shuttles, *Endeavour, Atlantis,* and *Discovery,* and move on to a new type of spacecraft called the Crew Exploration Vehicle or CEV. The final countdown for the shuttle program is tentatively set for the year 2010.

NASA's uncertainty over its new moon-and-Mars mission was a departure from what Americans who watched the space program were used to. Since 1981, shuttle astronauts have given a thumbs-up for the cameras as they paraded in their spacesuits from their crew quarters at the Kennedy Space Center to board the AstroVan. After the brief trip to the launchpad, they would squeeze into one of the five shuttles for liftoff. In 1984 when one of these spacecraft, known simply as an orbiter, deployed two satellites that failed to work properly, more astronauts were sent up. Two of them donned spacesuits and one used a Buck Rogers–style jet pack to retrieve the faulty spacecraft for repairs. When the shuttle launched the Hubble Space Telescope in 1991 with a misshapen main mirror, more spacewalkers later installed a set of corrective lenses to straighten the observatory's fuzzy vision. The work exemplified the can-do attitude of the astronauts, despite the shuttle's unfocused role in the U.S. space program. The American public watched these space missions with curiosity, but also with disbelief as fourteen astronauts were killed along the way during the *Challenger* explosion in 1986 and the *Columbia* accident in 2003.

The triumphs and tragedies of the shuttle (officially called Space Transportation System, or STS) began after the Apollo moon program was cancelled because of budget cuts in the 1960s and 1970s. Tighter funding left NASA relegated to work that was more modest, less defined, and closer to Earth than the moon voyages.[2] The strolls on the gray lunar landscape that started in 1969 were followed up with the Skylab space station and the Apollo-Soyuz Test Project in the 1970s, and ultimately the shuttle in the decades after that. Skylab gave nine astronauts, seven of whom were denied moon flights, the consolation prize of circling the globe for a few months. Apollo-Soyuz was the last gasp of the lunar program, featuring the first-ever docking of a U.S. spaceship with one from the Soviet Union. After that launch in 1975, NASA was grounded until the first shuttle flight in 1981. Following years of high adventure, even NASA acknowledges that the space shuttle was the wrong path to take, and the decisions that led to the vehicle's creation were flawed. That's a far cry from the space plane's previous position as the crown jewel of the U.S. manned space effort for almost three decades.[3]

During the shuttle program's inception in the late 1960s, proponents claimed the spacecraft would be useful in building a space station, by "shuttling" people and materials to and from orbit at low cost. The problem was that there wouldn't be support from Congress to start building a space station until almost twenty years after the maiden flight of the shuttle. This lack of a "big picture" purpose for the orbiter while it was still being designed left NASA looking for jobs for it to do, to justify the expense to U.S. taxpayers and the risk to the astronauts. The agency then employed salesmanship to enlist the military as a customer for the new craft to encourage Congressional funding.[4]

With no space station to serve, the shuttle was advertised as the future workhorse of the Pentagon to carry spy satellites to orbit. In reality, only a fraction of those secret payloads ever flew.[5] The new spaceship, NASA promised, would also launch commercial satellites for paying customers, even though unmanned rockets were judged to be safer and less expensive. Finally, the agency said that the winged vehicle would build America's expertise in space. But with no clear mission, what specific skills would be needed and how would those abilities be used?

NASA and the Pentagon each contributed to the final design of the space shuttle, and then the agency pushed the first orbiter, *Columbia*, onto the launchpad.[6] With no grandiose vision like Apollo, the craft would spend most of its early life on what would amount to little more than errands in space. There would be four test flights, starting in 1981, and then the vehicle would be declared operational. After that, on one trip to orbit the ship's list of tasks would include carrying and deploying commercial satellites. The next flight might feature a laboratory compartment installed in the shuttle's big cargo bay so the astronauts could float inside and conduct experiments, and then more satellites would be launched on the liftoff after that. NASA appeared to be going places, just nowhere in particular.

The program would find some purpose, late in life, by finally carrying up parts of the International Space Station. But that usefulness would be cut short one year after the *Columbia* accident in 2003, when the White House ordered shuttle flights to be terminated in the year 2010. Pursuing a new direction with the new Crew Exploration Vehicle, and its goal of venturing to the moon and Mars, may give NASA the chance to put the past behind it and get back on track with a true mission to fulfill.

Even before rocket scientist Michael Griffin was nominated by President Bush in 2005 to be NASA's top administrator and the standard-bearer for the White House's new space initiative, he helped to write a report for the Planetary Society that strongly supported the idea that NASA cut its losses, scrap the shuttle, and move on. In its place, the report recommended a new ship that could be configured to fly to the space station, make trips to the moon, or carry crews on long-duration flights to Mars to either orbit the planet or land, depending on the phase of ongoing exploration. The document was issued two months after Bush gave NASA its new marching orders.[7] Less than a year later, the president offered Griffin the top job at the U.S. space agency.

One month after his official confirmation in the U.S. Senate, when some might have taken a more conciliatory tone on Capitol Hill, Griffin's statements regarding the space shuttle remained harsh. "The shuttles are inherently flawed," he told a Senate subcommittee whose members represented states like Florida and Texas, where many NASA jobs hinged on the agency's future. "It does not have an escape system for the crew," Griffin added. "And we all know, since human perfection is unattainable, sooner or later there will be another shuttle accident." Instead of being a cheerleader for the shuttle, NASA's new boss considered the space plane an extremely aggressive design that was just barely able to work. This was an opinion he was ready to share, not only with members of Congress who controlled NASA's budget, but with the press as well. "I wouldn't characterize the shuttle as a mistake," Griffin told the BBC, "but the decision to limit ourselves to low earth orbit is not one that we [NASA and the Bush White House] would have made." Despite the more generous tone, Griffin's overall message was that he was there to bury the shuttle, not to praise it. It was also acceptance that he was leading a space program with problems rooted in history.[8]

Once the fleet of orbiters is gone, the new picture of NASA that the White House is painting is going to be a major change from the shuttle and nothing like the majesty of the huge Saturn V rockets that thundered off the launchpad in the 1960s to carry men to the moon. The next generation of astronauts is expected to blast off in a spacecraft perched on top of a single solid rocket booster from the space shuttle era. Assuming the program succeeds, it will be less spectacular to watch. But, unlike with the shuttle, NASA generally knows what it wants the new spacecraft to do.

Mr. Bush's 2004 announcement to end the shuttle program and aim higher, for the moon and Mars, wasn't the first time the U.S. space agency had been tasked from the Oval Office with an ambitious goal in space. The glory days of Apollo's manned moon landings were fueled by a previous challenge from a previous White House. Apollo was born while the Soviet Union was still basking in the glow of the successful launch of its Sputnik satellite and of cosmonaut Yuri Gagarin's voyage as the first man in space. Months later, on September 12, 1962, President John F. Kennedy made a speech at Rice University in Houston, Texas, that schoolchildren can recite to this day.

"We choose to go the moon in this decade and do the other things, not because they are easy, but because they are hard."[9]

What America likely didn't hear at that time was the sound of NASA gasping. President Kennedy issued his challenge after the flight of astronaut Alan Shepard as the first American in space in 1961. That mission meant a grand total of fifteen minutes of flight experience during a quick hop from Cape Canaveral on Florida's Atlantic coast to the Bahamas, with no time in orbit. Prior to that minor triumph, NASA engineers working at the fledgling Cape Canaveral rocket facility recalled tuning their radios to pick up the rhythmic beep-beep-beep as the basketball-sized Sputnik satellite and its battery-operated radio sailed over the United States.[10] Kennedy's lofty dream was followed by budget battles in Congress, lobbying among contractors who wanted to build the hardware, a little foreign competition, and uncertain career paths for the astronauts who fought for a seat on the spacecraft. The difference between Apollo and the space shuttle program is that President Kennedy chose the outcome, namely a moon landing, and NASA was put in the initially uncomfortable position of finding a way to achieve this. It would build a tailor-made spacecraft to accomplish its goal.

Bush's new moon-and-Mars program, which signaled the demise of the shuttle, may have prompted a feeling of déjà vu at NASA, especially among longtime employees with Apollo memorabilia on their office walls. There would be new and precise goals, a new debate over funding, competing spacecraft designs, and a fresh lineup of potential astronauts who would see their dreams succeed or fail miserably.

President George W. Bush Aims for the Moon and Mars

February of 2004 left NASA with a familiar knot in its stomach. President Bush had just ordered the new moon-and-Mars initiative, similar to President John F. Kennedy's challenge for the first Apollo moon landing.

"This cause of exploration and discovery is not an option we choose," said President Bush. "It is a desire written in the human heart."[11]

The speech was delivered to observe the first anniversary of the 2003 *Columbia* disaster. President Bush consoled the nation on that day. Now, one year after the accident, he set his sights on a new mission for NASA. His quest to the moon and Mars was framed in equally stirring words, but the goal left NASA with the same question it faced in the 1960s—how to do it.

The U.S. space program sacrificed the ability to fly to the moon when Apollo was phased out. The shuttle was built to fly only to low earth orbit. NASA also lacked substantial experience in the effects of prolonged spaceflight on the human body. Astronauts returning from weeklong shuttle missions experienced disorientation and showed early signs of loss of bone density and muscle mass. What would a long flight to Mars do to them? For the most part, NASA didn't know.

At the time of Mr. Bush's address, the U.S. endurance record was about six months in space. That was when astronaut Shannon Lucid lived and worked in orbit aboard the Russian space station Mir. She returned to Earth to a shower of colorful commemorative packets of M&Ms, which she had once confided to reporters was the snack item she missed the most during her half year in space. Depending on the flight path, a trip to the Red Planet might take as long as Lucid's trip on Mir, or longer. And that doesn't include the return flight, where the astronauts have to be kept alive and well until they land back on Earth.

Space Contractors Play "Let's Make a Deal" for the Crew Exploration Vehicle

The starter's gun had been fired on NASA's new mission in 2004, and within a year America's aerospace contractors were ready to start bargaining to build the spacecraft. On January 30, 2005, the contractors who built NASA's hardware, past and present, gathered at the Walt Disney World theme park for what the agency called its 1st Space Exploration Conference. It was a how-to gathering on achieving President Bush's goal of a moon-and-Mars flight. It might also be thought of as a dinner bell rung for another possible NASA gravy train. The official agenda included discussions about the crossroads at which NASA found itself following the *Columbia* disaster.

The participants convened at Walt Disney World's Contemporary Resort Hotel. The beehive-shaped structure represented Walt Disney's vision of futuristic architecture in the 1960s, even before 8-track tape players were considered high technology. Monorail trains zip through the center of the building, and its big windows afford a view of the Magic Kingdom, including Tomorrowland with attractions like Space Mountain and Starjets, where tourists line up to fly round and round in little spaceships of their own.

The Contemporary Resort is also the place where President Richard Nixon, before he resigned in the wake of Watergate, proclaimed to the American people that he was "not a crook." It was Nixon, not John F. Kennedy, who welcomed back astronauts Neil Armstrong, Buzz Aldrin, and Mike Collins from the first manned moon landing—an achievement that occurred eight years after JFK's audacious challenge to America to put a man on the lunar surface by 1970.[12]

Industry exhibitors set up booths to proudly illustrate past victories in the conquest of space and to lobby, not too subtly, for more business from Washington. The list of those coming to Orlando for the conference included Boeing, which built NASA's space shuttle fleet; Northrop Grumman, which built the buglike Apollo lunar modules that carried men to the surface of the moon; and Spacehab, which built reusable cargo modules for the space shuttle to resupply both the Russian space station Mir and the International Space Station. The "big guys" among the contractors, like Boeing and Lockheed, were veterans of the Apollo program, where Kennedy pointed to the moon and then ordered NASA to find a way there.[13]

Step one is the ship. The man in charge of developing the new vehicle is NASA's CEV program manager, Michael Hecker. When the starter's flag fell on the design competition for the new spacecraft, Hecker was very pragmatic about what the final craft might look like. "You can have a capsule, or something with wings for coming back. All we've told industry is we need something that goes from the ground to low lunar orbit and back again." Boeing and Lockheed Martin led the two main competing design teams to win NASA's CEV contract, and Lockheed wasted no time in tossing its proposal into the ring.

The new craft will need to achieve low earth orbit, like the shuttle, and be able to dock to the International Space Station, again like the shuttle. But the CEV will set its sights higher than its ancestor, for it must also be able to go to the moon or Mars. Not even the primary contractors vying to build the craft knew early on how that might happen. NASA's CEV Program Office says it doesn't care how the spacecraft finally looks, so long as it gets the job done. "The designs cover the gamut," says Mike Hecker. "We have ideas for craft that drop straight down"—like the Apollo capsules that plopped into the ocean after a mission—"or have wings for downrange flying. It's the gamut yet."[14]

In short, during the early stages, wings were possible and so was a blunt-bottomed capsule.

One thing that all members of the design competition appear to agree on is that the CEV will not use heat protection tiles. The mosaic of tiles pasted on the outside of the space shuttles *Columbia, Challenger, Discovery, Atlantis,* and *Endeavour* was developed to provide the spacecraft with reusable protection from the heat of reentry. It also made the orbiter vehicles look decidedly odd when you got up close. On the launchpad, each space shuttle appears sleek and clean with its orange-colored external fuel tank and gleaming solid rocket boosters. Upon closer inspection, however, the shuttles take on a "tattooed" appearance.

The tiles are all marked with serial numbers indicating their location on the surface of the orbiter. Each custom-made tile is cut specifically to fit around a wing, for instance, or next to a cockpit windshield, so the shuttles are covered with tiny black or white numbers. These actually came in handy following the *Columbia* and *Challenger* disasters. As search teams found pieces of wreckage with the tiles and serial numbers intact, NASA engineers compared the numbers to a chart to learn the exact spot where a jagged piece of debris belonged on the orbiter before it was destroyed.[15]

So don't look for tiles on the CEV; they are not likely to be there. The two semifinalists in the fight to win the CEV contract, Northrop and Lockheed Martin, both appeared interested in the heavy-duty reinforced carbon-carbon shields made famous by the *Columbia* accident. These slate-colored shields are tougher than the lightweight ceramic tiles.[16] A falling piece of orange foam insulation from *Columbia*'s external fuel tank broke a shield on the leading edge of the left-hand wing. That allowed air, heated to 3,000 degrees Fahrenheit by the friction of reentry into the earth's atmosphere, to melt the spacecraft's aluminum frame. The vehicle burned up and the seven astronauts aboard died. Still, the heavy shields were more robust than the lighter ceramic heat tiles, which could be pierced by the point of a pencil.

The subject of crew safety also prompted members of the NASA astronaut corps to be especially vocal. Shuttle veterans have long complained about the lack of a crew escape system on their spacecraft. Launching humans with solid-fuel boosters has always been controversial, since that kind of rocket can't be shut off if there's a problem. Liquid-fueled engines can be throttled back, but once solid fuel is ignited, it burns until the propellant is exhausted. That was one of many uncomfortable compromises made to get the shuttle program moving. A leading proponent of a crew escape system for the new CEV is NASA's first female space shuttle commander, Eileen Collins. Her first flight was with cosmonaut Vladimir Titov, who had a close call on an earlier Soyuz rocket launch. "There was a launchpad fire," says Collins. "And Titov and his crewmate ejected safely. In fact, they were up and walking around shortly afterward." Early designs for the CEV include a launch escape rocket on top, which could pull the craft away in case of an emergency.[17] Of course, this will be a benefit for future crews and not NASA's current list of astronauts.

While a crew escape system was missing from the shuttle, the new CEV will be lacking one component from the days of Apollo, even though its purpose is also to make lunar missions. There will be no pricey rocket like the Saturn V to send the new spacecraft to the moon. Instead, smaller boosters that are currently in use to launch unmanned satellites were considered for CEV missions to the moon or Mars. That scenario meant using small rockets to carry up crew members or components to outfit a CEV for a trip to lunar orbit or the long haul to Mars. In other words, the concept was to send up astronauts on rockets originally designed to launch satellites. The Saturn rockets that carried astronauts to the moon were so powerful that they rattled the windows of homes all the way from the Atlantic coast to landlocked Orlando. Space shuttle blastoffs have been more

Figure 1. Artist's conception of the launch of the Orion space capsule propelled into orbit by NASA's new Ares rocket. The spacecraft is the replacement for the space shuttle fleet, and can carry up to six astronauts. The new rocket is similar to the solid rocket boosters used on the shuttle. The Ares has five solid rocket segments stacked one atop the other, instead of the four of the shuttle era. Courtesy of NASA/Lockheed Martin.

modest, each generating barely enough vibrations to prompt reporters at the Cape to lift their microphone stands off their desks. The shockwaves from the CEV launches are expected to be smaller still.

Michael Griffin's Vision of the CEV

At the space exploration conference at Disney World, participants proposed a number of ways to build the new spacecraft and launch it. On September 19, 2005, NASA administrator Michael Griffin dashed the hopes of contractors who wanted to sell expendable rockets like the Delta-4 or Atlas-5 to put the next generation of Armstrongs and Aldrins on the moon. Instead Griffin unveiled his own ideas for the revamped lunar program, which envisioned the first return missions to the moon blasting off on two rockets, not one. The skinny, needlelike crew rocket would fly only after a bigger cargo craft built with shuttle technology was launched into space first.[18] This cargo rocket, powered by the bigger shuttle boosters and a taller version of the shuttle external fuel tank, would carry a buglike lunar lander. The two ships would dock, and then head to the moon for expeditions lasting about a week. "Call it Apollo on steroids," says Griffin. Four astronauts would go to the moon, compared to the crews of three during Apollo. The weeklong excursions on the lunar surface would top Apollo threefold, and at only half the cost, according to NASA number crunchers. Reporters were quick to pounce on the estimated $104 billion price tag. Griffin contended that the program could proceed without additional funding from Congress.[19]

Politics and budgets notwithstanding, perhaps the biggest irony in NASA's moon plan is that it appeared to take a page from the Russian playbook. The early Soviet plan to beat the Americans to the lunar surface involved launching pieces of its Soyuz moonship one rocket at a time. It also built on previous spacecraft designs instead of tossing out the old ones to bring in the new.[20]

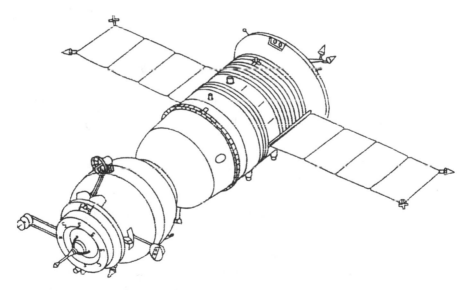

Figure 2. Soyuz TM. The Soyuz spacecraft began taking up cosmonauts in 1967. The first flight ended disastrously with a crash landing and the death of the pilot. The craft later became the backbone of the Russian space program. It also temporarily replaced the U.S. space shuttle following the 2003 *Columbia* accident, delivering crews to the International Space Station. Courtesy of NASA.

The current model of Soyuz (meaning "union") can trace the vehicle's lineage all the way back to the first one, which flew in 1967 with tragic results. The capsule's electronics failed, and then its parachute lines snarled during the plunge to Earth. The one cosmonaut on board was killed.

From Griffin's perspective, the piecemeal strategy to reach the moon also carries other pluses. One, it could spare the NASA workforce a jarring transition as the nation trades the shuttle for a completely new way to get into space. Official estimates from the Kennedy Space Center that the workforce that has maintained the shuttles would plummet following the retirement of the fleet rang alarm bells in the communities surrounding the Cape. Proponents of shuttle-derived designs believe that more jobs might be salvaged as America moves from the orbiters to the CEV. Understandably, members of Congress in areas that benefit from the space program were quick to sign on to Griffin's plan.

In Brevard County, home of the Kennedy Space Center, the days following *Apollo 11* are vividly recalled for their negative impact. Fewer launches meant fewer people were needed to build and assemble the big Saturn V rockets. NASA's own workers kept their jobs as the nation moved slowly toward launching the new space shuttle in 1981. NASA contract workers didn't fare as well. Thousands were left without jobs, or forced to move away to work on projects not located on Florida's Space Coast. Word that the shuttles were about to retire rekindled memories of the post-Apollo depression in Brevard. Utilizing shuttle technology, as Michael Griffin suggests, might help keep some of KSC's engineers on the payroll.[21]

Lockheed Martin Catches the Brass Ring

The months leading up the final selection of the contractor to design and build the new Crew Exploration Vehicle capsule, and the rocket that would carry it, saw a combination of political maneuvering, high technology, and show biz. Lockheed staged a press event at the headquarters of the Florida Spaceport Authority to publicize the benefits Central Florida would see, assuming Lockheed won the lucrative CEV contract from NASA. The authority's job is to foster development of space-related business in Florida, and that was clearly on the agenda that day. Top company managers were joined by local members of the Economic Development

Authority and Florida's lieutenant governor, Toni Jennings, to announce a proposed partnership including a $45 million financial incentive package from the state. The group sat on a stage, next to a large scale model of the new spacecraft, to field questions from reporters. The deal meant Lockheed would assemble and test its CEV capsule at the Kennedy Space Center, generating a possible two to three hundred new jobs in an area still nervous about the end of the shuttle program. All of this bounty for the community depended, of course, on Lockheed winning the deal. The State of Florida offered a similar incentive package to the other competitor in the race to build the CEV, Northrop-Grumman-Boeing, just to hedge its bets. "We're not playing favorites," said Lieutenant Governor Jennings.[22] As to where Northrop might assemble its capsule, assuming the decision went the other way, that company was keeping its cards face down for the time being.

Attendees at the Lockheed event were invited to go home with a black ceramic coffee mug bearing the logo of Lockheed Martin's CEV capsule proudly circling the moon. Just as a defense attorney might deliver his final arguments and then clench a cigar triumphantly in his teeth and strut out of the courtroom, the event was designed to depict Lockheed as "the winner" long before the final announcement was made.[23]

August 31, 2006, was the day NASA planned to announce the name of the contractor, Lockheed or Northrop, that had won the right to build the successor to the space shuttle. The grand event would also unveil the name of the capsule that would carry the agency into the twenty-first century. In an effort not to leave the International Space Station out of the excitement, NASA carefully arranged for astronaut Jeff Williams to record a video message from inside the orbiting complex to help unveil the selection of the capsule's name. "We've been calling it the Crew Exploration Vehicle for several years," Williams said. "But today it has a name, Orion." The catch was, the message was heard by members of the press over air-to-ground radio on August 22, not the end of the month when NASA intended to make the news public. Headlines blazed the name *Orion* within hours.[24] On August 31, with some of its thunder gone, NASA announced that Lockheed had won the deal. It still meant an estimated $8 billion for the contractor, and that was just for the early phases. The announcement also started a flurry of work between the company and the people who would fly the vehicle, NASA's astronaut corps.

Figure 3. Artist's rendering of NASA's new Orion capsule, also known as the Crew Exploration Vehicle or CEV, in orbit around the moon. The lunar landing vehicle (not shown) will be piloted to the surface below for missions longer than the Apollo flights in the 1960s and 1970s. Courtesy of NASA/Lockheed Martin.

Developing New Rockets along with New Capsules

Amid the embarrassment over the naming gaffe, NASA actually had two new names to deal with as it moved into the new era of the Crew Exploration Vehicle. One was Orion, for the crew capsule, and the other was Ares, the name for the two distinct types of rockets that would send both people and cargo into earth orbit on missions to the International Space Station and, perhaps, to the moon or Mars. Both were direct descendants of space shuttle technology, and the early artists' conceptions and plans for unmanned test flights hammered that notion home. NASA intended a number of launches to confirm the soundness of the Ares design before allowing it to carry humans aloft. One flight would utilize hardware left over from the first space shuttle liftoff following the 2003 *Columbia* accident. Shuttle *Discovery*, on mission STS-114, had two sets of solid rocket boosters. One pair had been replaced before liftoff by a second set. Four cylindrical segments of one of the discarded boosters were selected for use in a test launch of the Ares-1, or the rocket that would propel astronauts inside the CEV capsule. The actual rocket would have five segments, not the four used by the shuttle and the planned test vehicle. Still, NASA felt that with a ballasted upper stage built to mimic the size and weight of the fully loaded Orion capsule and its J2X second-stage booster, engineers would get a good enough notion of the vehicle's performance to help with the final design of the new spacecraft and the rocket. Similar tests were planned for the Ares-5 cargo rocket that would eventually carry the lunar lander for trips to the moon's surface.[25]

While all these plans were laid out, fans of the *Star Trek* television series and movies might have had a chuckle over the logo for the Orion capsule stitched onto many of the polo shirts worn by key managers and technicians. The symbolic name *Orion* with a blue Planet Earth in place of the capital letter O had a red orbital loop for the path from our planet toward three stars representing the three destinations of moon, Mars, and beyond. The design had its origin in the world of science fiction, since it had been created by Michael Okuda, an artist who had come up with many of the sets, props, and spaceships used in the *Star Trek* television series and films.[26]

How Many Windows, and Where Does the Toilet Go?

While NASA's engineering community dealt with test shots of rockets, the astronaut corps at the Johnson Space Center had more basic concerns. The astronauts worked from the start to make their voices heard regarding the new craft that would replace the shuttle. Even before the White House announcement that the shuttle fleet would be retired, there had been lobbying for a crew escape system on whatever was selected to be the next ship. Now the astronauts were invited to take part in designing the inside of the Orion capsule, and veteran spacewalker Lee Morin would get to utilize his experience designing interactive cockpit equipment for fighter jets.

Morin had only one shuttle flight to his credit, but it was a busy one. During the STS-110 mission in the year 2002 he performed two space-walks, totaling fourteen hours, to help install the first major section of the International Space Station's long, spinelike exterior frame. Now he was deputy manager of the Astronaut Office's Orion Cockpit Task Group. Perhaps the first thing NASA's veteran crew members would have to accept was that the new CEV was going to be different from the shuttle in almost every way.

Difference number one will likely be less elbow room. The shuttle has a two-story crew cabin, with the cockpit for the pilots on the upper deck, and a lower floor called the middeck for eating, sleeping, and conducting scientific experiments. For the four to six crew members of Orion, each trip in the gumdrop-shaped capsule will seem like moving to a one-room studio from a spacious one-bedroom apartment. "The shuttle's got the flight deck, the middeck, the airlock, and the tunnel going to the space station," Morin says. "The CEV is going to have none of that. It's going to be cramped." His flight to the station in 2002 meant docking *Atlantis* to the orbiting complex and unloading the big truss section. His spacewalks all began by exiting the outpost's airlock instead of the one on the shuttle. Despite the work that required Morin's crew to be inside the compartments of the space station, they found themselves spending as much time aboard *Atlantis*. "We slept on the shuttle, we ate on the shuttle, we used the toilet on the shuttle," he says. "I spent one night sleeping on the station just to say I did." By contrast, astronauts arriving at the station aboard the CEV might be highly motivated to get out of the capsule and into the more spacious modules of ISS, if for no other reason than to avoid bumping into their crewmates in the small spacecraft.[27]

Designers of the space shuttle got around the lack of space inside the crew cabin by making most of the seats removable. The only flight couches that are permanently bolted to the floor are for the commander and the pilot, in front of the main cockpit windows. Once in orbit, the rest of the crew members can unstrap themselves, disconnect the chairs they sat in during liftoff, and fold them for storage. Each seat fits into a compact locker space no bigger than two cubic feet.[28] Builders of the new CEV crew capsule won't have that option. The seats on Orion have to be permanently installed and ready to be occupied, because the capsule will be the escape craft, or "lifeboat," for the International Space Station. If the

Figure 4. Artist's conception of the Orion spacecraft on final approach to the International Space Station. The new crew capsule is intended initially to visit the space station to deliver visiting crews and act as the orbiting outpost's emergency lifeboat. After test runs to ISS, future missions are meant to return astronauts to the moon and, perhaps, fly on to Mars. Courtesy of NASA/Lockheed Martin.

crew of the orbiting complex need to abandon ship, they may have only minutes to depart, depending on the emergency. "If you puncture a crew module or something, you have to get in your spacesuits and get out of there," says astronaut Lee Morin. "You won't have time to deal with the Legos"—the folding flight seats. Beneath the flight couches on the CEV will be storage lockers similar to the ones on the shuttle. When it comes to making use of limited space, NASA may look to the Russian space program and its tightly packed Soyuz crew capsules for inspiration on how to fit in as much gear as possible. Astronauts like Norman Thagard, Mike Fincke, and Ed Lu have made their trips to the International Space Station in Soyuz vehicles, strapped into seats with their knees pulled up almost to their chests, with bulky gear-filled duffel bags stowed just inches away. "You use all the nooks and crannies in a Soyuz," says Lee Morin. "You see things like clothing you need for survival situations stuffed everywhere." Beneath the CEV flight seats there could be a flexible grid, corresponding to the floor, with locker space below that.[29]

Beyond the configuration of the Orion craft, NASA will likely have to deal with the spacesuits the CEV crew will wear. The Astronaut Office believes the orange pressure suits and helmets worn on the shuttle won't be adequate for the new capsule. The shuttle suits have been called "launch and entry suits" or "high altitude protection suits" over the years. Each was designed, not for floating in outer space, but for the bail-out survival option shortly after liftoff and just before landing. The ninety-pound suits are loaded down with a parachute, a survival raft for water landings, dye markers to signal to rescue ships or helicopters, and drinking water, among other items. Shuttle spacewalkers wear much heavier suits and float out from an airlock while their crewmates remain in the comfortable shirt-sleeve environment of the orbiter.[30] Again, the CEV capsule won't have that option. "If one crewman does a spacewalk, they all do," says Morin. That's because there's no room for an airlock in the new spacecraft. Any problem that requires going outside, like a damaged solar panel, means depressurizing the whole vehicle and opening the hatch. If that remains the case, NASA will need to design suits robust enough to stand up to the hostile conditions of space, yet practical in size so everyone can fit inside the capsule.

According to early designs for Orion's cockpit, crew members that squeeze inside for liftoff will also see a much less complicated interior than the space shuttle's. The inside of the orbiter vehicle has almost no empty wall space. Some of the bulkheads are crammed with control panels covered with switches, dials, gauges, or brightly colored computer screens. Below, storage lockers occupy much of the middeck, along with the toilet, and the airlock for spacewalks. The collection of buttons and dials that shuttle astronauts have to keep track of totals around two thousand. NASA estimates the cockpit of the new CEV capsule will contain a fraction of that, maybe a hundred switches at most. Replacing the gadget-filled shuttle layout there might be one central control panel with a few touch-sensitive computer screens, each surrounded by large buttons for crew use. Another expected difference is no checklists or books floating around the cabin, since Orion is being designed to be "paperless." Instead of printed documentation to instruct the crew and keep the mission organized, NASA is thinking of giving each crew member an electronic organizer called a "portable electronic appliance." It's not quite a laptop, or a Palm Pilot, or a Blackberry. "Think of it as like an Etch A Sketch," Morin says. In flight the devices would be stored in lockers beneath the seats. "There might be a hardbound book on how to reboot things," the astronaut adds, "but that's about it." The goal is to avoid limiting the new spacecraft to today's technology. The space shuttle was criticized for having a cockpit stuck in the 1970s with mechanical gauges and switches. The craft had been flying for more than a decade before NASA updated the dashboard with the brightly colored computer screens common on commercial passenger jets. The new CEV is meant to be equipped with whatever state-of-the-art computer hardware is available.

The Astronaut Office in Houston is working to have its say in almost every aspect of the new CEV, from high-tech matters to more basic things like the toilet. That's a question that astronauts say they often get from schoolchildren, but how the "facilities" are to be included on the Orion capsule has already created disagreements between astronauts and engineers. "One designer came up and said he wanted to put the toilet in the middle of the capsule," says Morin. "We said no."[31]

The New View from the Top

The windows that will go on the craft are another source of discussion. Whereas the space shuttle had eleven windshields or portholes, the CEV may have no more than four, so future crew members might have to compete with each other for a chance to look out much smaller and fewer windows.

Initial designs for the CEV spacecraft include two primary windows, one each for the commander and the pilot, and perhaps a couple of smaller portholes. Members of the Astronaut Office took the early blueprints of the capsule's windows to a room called the Reconfigurable Orbital Cockpit facility, where computers were used to cut Styrofoam mock-ups of the window frames. These facsimile windows were then installed in a room that has equipment to project on the wall images of the earth or an approaching space station. That way, veteran shuttle pilots could look

Figure 5. Early visualization of Orion in lunar orbit preparing to make NASA's first moon landing since *Apollo 17* in 1972. NASA estimates a crew of four astronauts would make the voyage instead of the three people who flew each mission during Project Apollo. Courtesy of NASA/Lockheed Martin.

through the foam window frames and judge what their field of vision might be like in space on the new ship. Even with the give-and-take between astronauts and Lockheed designers, longtime crew members admit the windows must be smaller than those in front of the commander and pilot on the shuttle's cockpit deck, six king-sized windshields that measured forty-two inches diagonally. The CEV is more likely to have windows the size of the ones at the aft end of the flight deck, some twenty inches on a side.[32]

What may aggravate crew members of the space shuttle generation the most is that the CEV will require less actual piloting than the outgoing winged space plane. The shuttles reenter the atmosphere in similar fashion to the blunt-end Apollo capsules of the 1960s. But in the final minutes before landing, the commander and pilot get to take the stick and guide the craft to touchdown, using the wings and tail rudder like a true airplane. The CEV will "plop" back to Earth by parachute, on land or, like its Apollo-age predecessors, into the ocean. The prospect of the new capsule affording less "righteous" touchdowns for the CEV pilots draws dissent from the astronauts. They point to abort modes during launch, docking, or landing that require action from the commander and the pilot. Then there are the proposed missions to return to the lunar surface. "If you want the feeling of the wind whistling through your hair," says Lee Morin, "try landing on the moon." And then there are the possible voyages to the surface of Mars, assuming NASA is allowed to go that far.[33]

The Pros and Cons of Going "Apollo Style"

The plan for returning astronauts to the moon with an eye on Mars would have to be made step by step—the same way NASA did it during the race to the moon in the 1960s.

Apollo 11 command module pilot Michael Collins once described the process of going to the moon as a "daisy chain" of steps leading up to a lunar landing. In other words, it consisted of solving problems one at a time along the way. For Apollo, that meant first making the simple orbital flights of Mercury with pioneers like John Glenn, Alan Shepard, and Scott Carpenter. These missions confirmed that human beings could survive the rigors of weightlessness. Visitors to the Smithsonian Air and Space Museum can see the small eye charts inside the *Friendship 7* Mercury capsule that John Glenn occupied during his three orbits of the earth. NASA physicians weren't sure if a man's eyes would change shape, making the piloting of a spacecraft impossible.

These first flights were followed up by the two-person Gemini missions that proved people could pilot spaceships, change orbit, rendezvous with other vehicles, and dock with those craft. These skills would prove useful as Apollo crewmen flew to the moon in two vehicles—the mother ship, or command module, and the lunar module designed to take two astronauts to the cratered surface of the moon. That piecemeal process enabled NASA to fulfill President Kennedy's goal of putting people on the moon.[34]

NASA basically received a blank check in the 1960s to achieve that goal, which meant beating the Russians and getting to space fast. Debate raged in Washington on how to do it, and the consensus was to abandon rocket planes like North American's X-15. This cutting-edge vehicle, after being launched from the underside of a B-52 jet bomber, would fire its own engines to fly to the edge of space. But compared with the stubby-winged X-15, rockets and capsules seemed a better bet. The one-man Mercury capsules were first perched on Redstone rockets, which were originally designed to carry nuclear weapons. "Small" was the watchword. In fact, Mercury pioneer John Glenn once said you don't climb into a Mercury capsule, "you put it on."[35]

Glenn flew aboard the space shuttle in 1998 and spent most of his time in orbit conducting experiments comparing the effects of age to the effects of weightlessness on the human body. The medical objectives of his Mercury mission in 1962 were much more basic. Those missions helped answer simple questions about how humans functioned in the absence of gravity. "NASA didn't know if fluid would float in your inner ear," Glenn says. "The result of that could be uncontrolled nausea." The early Mercury flights proved that didn't happen, although space sickness would later strike about one in three astronauts, and there was no telling who would be stricken.[36]

Next there were the two-man Gemini flights. It was Alex Nagy of NASA headquarters who thought of naming the program Gemini after the celestial twins. These next-generation spacecraft, originally known as Mercury Mark-2, proved that two spacecraft could rendezvous in orbit. That was a skill necessary for the lunar-orbit docking scenario NASA favored for the Apollo landings. Gemini missions also proved that onboard computers would work in space, and that maneuvering thrusters using two self-igniting chemicals were a practical way for a craft to change altitude, position, or orbit as the mission required. These flights demonstrated, too, that people could work for long periods in space.

Finally, the Apollo missions took astronauts to the moon on the first rockets designed specifically to carry human beings. But with success came a more blasé attitude about space. When the final manned moon mission was under way in 1972, longtime NBC space reporter Jay Barbree caught himself saying something that would have been unheard of just a couple of years before. An editor called to ask what was going on with the astronauts aboard *Apollo 17*.

"Oh, nothing," Barbree replied. "They're just going around the moon again."[37]

Not even trudging around the surface of the moon could hold the attention of the American people, let alone pry open the checkbook of the U.S. Congress on a continuous basis. One reason was the primary goal that Apollo was supposed to achieve, namely to put a man on the moon before the Soviet Union. After that, the world asked, what next? Things would become even more routine following NASA's decision, prompted by budgetary prodding from Congress, to trade big rockets and a specific mission for the winged space shuttle. The new vehicle was billed as NASA's "space truck," and the public found the notion of astronauts as moving-van drivers much less exciting than space pioneers battling the Soviet Union for supremacy among the stars. Project Constellation is framed to go to the moon and Mars as part of an ongoing mission of exploration, instead of making just "one small step" like Neil Armstrong's in 1969. NASA is already taking steps to try to keep it going.

Administrator Michael Griffin recognized, soon after taking the reins at NASA, how Apollo had fallen out of favor in the 1960s and that his new space initiative could suffer a similar fate. On May 18, 2005, he shared with members of the U.S. Senate Subcommittee on Science and Space his concerns about whether the new moon-and-Mars program could remain viable over the long term. "I do not want, and we should not want," he said, "to repeat the mistakes of the Apollo program, where many unique capabilities were shut down abruptly and irretrievably." The agency's original post-Apollo plan called for extended stays on the moon, space stations, and an ambitious expedition to the planet Mars. Future destinations like that were scratched from NASA's agenda as too costly.[38] The hard lesson for the U.S. space program, then and now, is that getting a new spacecraft and mission is one thing, but keeping both alive is another. It will largely depend on whether future members of Congress are willing to pay what could be a hefty price. Estimates vary, but that final cost would be in the hundreds of billions of dollars.

Critics Oppose Another Apollo-Style Budget Boondoggle

Not everyone is welcoming this grand new space challenge with open arms. The list of critics includes Dr. John Gibbons, science advisor to President Bill Clinton from 1992 to 1998. Gibbon was also hired by President Nixon in 1973 to be the first director of the Office of Energy Conservation. His criticism isn't limited to the current Bush White House but extends to the previous one, from 1988 to 1992. "George H. W. Bush talked about going to Mars, and you saw how fast that died," says Gibbons. There's a small, but vocal, group of scientists who believe that spending vast amounts of money on sending astronauts on manned missions is foolish, and Gibbons is among them. "There's been this urge by NASA, and the public I suppose, to send people out into space," says Gibbons. "And I think anyone who knows anything about astronomy would say that's mad." His contention is that robotic missions can do a better exploring job than astronauts. Along with Dr. Gibbons's harsh assessment, another shock to the system could be congressional unwillingness to fully bankroll the CEV.

NASA is already being criticized for becoming an agency with just one mission—putting people on the moon and Mars. That's a big reason why Dr. Gibbons dislikes Project Constellation so much. These views came in sharp contrast to Gibbon's battles in Congress to salvage the International Space Station. The orbiting outpost had foundered over an expensive decade or so with no hardware in orbit—that is, until Gibbons and the Clinton White House chose to champion the cause of building the complex. The station was seen as a way for nations around the globe to work together on a nonmilitary project.

But that doesn't mean Dr. Gibbons loved all things NASA. He remains a major proponent of using unmanned missions instead of sending astronauts. His stance meant taking political heat during the Clinton administration as NASA prepared to launch the robotic spacecraft called Cassini. The mission to Saturn utilized a thirty-foot-tall probe resembling a high-tech birdbath covered in gold foil. It was powered by radioisotope thermoelectric generators, or RTGs. Antinuclear activists loudly protested the plutonium that made the generators work. Still, Gibbons says unmanned missions are worth fighting for. "I believe in going to Mars," he says. "But we should go robotically, to send our senses there, just not our bodies. This sending people to Mars stuff is a dreamlike plan to escape the realities of Earth." Unfortunately, proponents of emphasizing science over heroism frequently run into a political reality in Washington: the axiom that "Congress is a people business" and getting your picture taken next to a space probe doesn't look as good as a "happy snap" with an astronaut just back from space.[39] With the arrival of the new Crew Exploration Vehicle and the new and focused mission of returning to the moon and heading on to Mars, a different collection of astronauts will be posing for those pictures. Many of the people who currently fly aboard the shuttle know that their days are numbered at NASA. The end of the space shuttle program means the end of them as well.

2

New Spaceship, New Astronauts

Retiring the space shuttle program may give NASA a chance to forget its troubled past and pursue the genuine mission it has lacked for years. At the same time, many of the men and women who spent time looking down on Earth through the double-paned windows of the shuttle orbiters are now sounding nostalgic, as well as resigned to the coming change.

In February of 2006, as astronaut Steve Lindsey and his crewmates were training for the second shuttle flight following the *Columbia* accident, they took time to meet with reporters at the Kennedy Space Center in Florida. Only six of the astronauts were present, since German crew member Thomas Reiter was training elsewhere for his long-duration flight aboard the International Space Station, which would begin after the shuttle undocked and departed. Lindsey, along with crewmates Mark Kelly, Piers Sellers, Lisa Nowak, Mike Fossum, and Stephanie Wilson gathered behind the podium in the Press Site auditorium to field questions. There were the usual inquiries about their upcoming flight, and concerns over last-minute wind tunnel tests to confirm the safety of modifications to the newly redesigned external fuel tank. Then a reporter asked the astronauts something that appeared to catch them off guard. "Are any of you thinking of training for a moon mission?" After exchanging looks with the other members of his crew, Steve Lindsey finally responded, "I think we're too old."[1]

In other words, as NASA pushes ahead to a new vehicle, Lindsey and his crewmates are passing the torch to a new generation. Some veterans of the Apollo program like Vance Brand, Ken Mattingly, and John Young remained at NASA following the demise of the moon program in 1975. They would find new chances to fly into space on the shuttle, and some current shuttle crew members may stick around until NASA returns to the moon. Still, there will be lots of new faces in the astronaut corps.

The years of the space shuttle were relatively comfortable ones for NASA and the American public, where its place in the world of space exploration was concerned. The United States enjoyed a leading role in space with regular trips carrying dozens of astronauts. With the exception of the Russians, anyone who wanted to go into orbit had to come to NASA "hat in hand" and ask for a seat on the shuttle. But it wasn't always like that.

During the dawn of the space race marked by astronauts like John Glenn and Alan Shepard, NASA was decidedly earthbound, and the Soviet Union was grabbing the headlines with success after success in space. In 1957 a Russian rocket carried up the first artificial satellite, called Sputnik. A follow-up launch featured a small pressurized capsule containing a dog named Laika. The press referred to that mission as Muttnik.[2] But it was when a slight man with a broad smile and the letters CCCP stenciled across his white space helmet first circled the earth in 1961 that NASA knew it had a true fight on its hands. Russian cosmonaut Yuri Gagarin became the first man in space, while unflown U.S. heroes like astronauts Gus Grissom and Gordon Cooper stood watching with their engineering colleagues at NASA who were struggling to keep their rockets from blowing up during one unmanned test after another. They were tasked with going to the moon, but that would be tough if they couldn't blast off without incinerating the people inside the space capsules.[3]

The World Reaction to Yuri Gagarin

In the early 1960s, while NASA personnel swatted mosquitoes, working without air conditioning in isolated Cape Canaveral, Florida, the Soviet Union was basking in the limelight. It was international attention sparked by a single phrase:

The man is in space!

That's how seventy-six-year-old Zoya Zarubina says she heard the news that the Russians had beaten the Americans into orbit with the flight of Yuri Gagarin. The white-haired, grandmotherly woman was, back at the time of Gagarin's historic flight, a chief translator in Moscow who would later work with the world's first cosmonaut. The Soviet government kept news of the blastoff secret until success was assured. "I was giving a lecture

Figure 6. Soviet cosmonaut Yuri Gagarin on his way to the launchpad for the liftoff of *Vostok 1*. His single orbit of the earth would prompt space enthusiasts to refer to him as the Columbus of the twentieth century. Courtesy of NASA.

before three hundred students, and my secretary had the audacity to come barging in during my talk," Zarubina says with mock indignation. "And I asked her what was the problem, and she said, 'The man is in space!'"[4]

In 1996 Zarubina was invited to the Kennedy Space Center as a guest to witness the liftoff of shuttle *Endeavour* on a mission to scan the earth with a wide, flat radar antenna nestled in the spacecraft's cargo bay. As interesting as the liftoff might have been, Zarubina seemed happier talking about Gagarin's mission in 1961 and her unique vantage point to watch the world's reaction. Back then, she was assigned to accompany Gagarin to Great Britain so he could be the drawing card at a Soviet exhibition in London. The world's first man in space was put in a borrowed Jaguar for the trip from Heathrow Airport to the Soviet embassy and then to the site of the exhibition.

Zarubina says both she and Gagarin were shocked by the size of the crowds that gathered at the airport to greet him, and the throngs of people who lined the streets to cheer the cosmonaut for his achievement. Zarubina recalled one small boy outfitted in a white helmet and orange jumpsuit, like the uniform Gagarin wore while squeezed inside the small Vostok capsule. Gagarin was delighted to see the young man dressed up for the occasion and invited him to sit in the car with him. The Soviet delegation was puzzled when the boy looked away each time a photographer tried to take a picture of him with Gagarin. Zarubina translated a question from the cosmonaut to find out what the problem was. "The boy said his sister worked all night making the costume for him," Zarubina recalled. "And he said he skipped class to see Gagarin and he was afraid his teachers would see him."

At a press conference, one reporter asked Gagarin whether he would accept an invitation to lunch with Queen Elizabeth, if it were offered. Gagarin responded that if a lady were to extend such an invitation, he would say yes. Shortly thereafter, an actual invitation came from Buckingham Palace. That prompted the cosmonaut to undergo additional training at the Soviet embassy on how to handle the complicated place setting with all the knives and forks that would surround his plate at the royal repast. Gagarin thought one fork was enough.[5]

There would be no luncheon with the queen of England for NASA at that time. Instead the agency fretted over how to catch up with Russia's string of successes: Gagarin's flight on *Vostok-1*, the first daylong flight by a Russian, the first spacewalk by a Russian, the first mission by a female Russian crew member. NASA would triumph in the end by learning along the way during the days of Apollo, as it did later with the shuttle. Now new hurdles have been set, with the moon and Mars as the goal of ongoing exploration.

Who Gets to Go?

The end of the shuttle and the start of the new moon-and-Mars program will represent a jarring change for the people who like to live and work in space. The CEV will again raise the question of who gets to go and who doesn't. Becoming an astronaut for NASA is a long process, with many successful space veterans later admitting that they tried and failed several times to join the agency before being accepted.

There is understandable concern in NASA's astronaut corps that, with shuttle missions winding down, there will be fewer and fewer seats available on remaining launches. Greater yet is the uncertainty about available places on the new CEV. A recurring topic of speculation at NASA is how many crew members might be able to fly on the new craft, with estimates ranging from four to six, instead of the seven or eight on space shuttle missions.[6]

Dealing with an uncertain future isn't new for astronauts. That goes back to the golden age of the program, when the moon seemed far away.

While America was gearing up for the space race in the 1960s, the U.S. Air Force was thinking of branching out into space on its own. It created a project called the Manned Orbiting Laboratory, or MOL for short. This was supposed to be the first covert space station, designed to eavesdrop on countries such as the Soviet Union. The MOL plan also opened the door to some would-be space travelers, including a navy test pilot named Robert Crippen. He would later copilot the first mission of the space shuttle in 1981 and command three later flights of his own. In the late sixties he was part of a group dreaming of spaceflight and looking for a way in. When the Pentagon announced that new astronauts were being sought, Crippen signed up and made it through the preliminaries. Then there was a fork in the road. Crippen needed to pick between NASA's civilian moon program or the clandestine military MOL program. "We looked at NASA," Crippen says, "and they had more astronauts they knew what to do with.

Figure 7. *Apollo 11* astronaut Edwin "Buzz" Aldrin during a fit check of his spacesuit prior to launch of the world's first manned moon landing. Courtesy of NASA.

So we thought this [MOL program] might be the way to go." Crippen was a test pilot instructor at Edwards Air Force Base in California when the MOL program started taking applications for prospective astronauts. He applied, and in 1966 he was selected to train for flight.[7]

The Manned Orbiting Laboratory built upon the Gemini program that paved the way to the moon. The two-man NASA capsules fine-tuned the agency's ability to conduct rendezvous and docking and acted as a platform for astronauts to practice spacewalking, where a crew member floated out of the capsule in a spacesuit. When NASA selected the *Apollo 11* crew to make the first moon landing—Neil Armstrong, Buzz Aldrin, and Michael Collins—they were all Gemini veterans.

The air force had something else in mind. Plans for the Manned Orbiting Laboratory meant hollowing out the upper stage of a Titan rocket and making the compartment into a two-man space station packed with surveillance equipment. The crew members would ride to space inside a Gemini capsule of their own, which remained connected to the rocket. The team would then float into their observation module and snoop on targets for the Pentagon. Crippen joined the original group of seven MOL astronauts that had been picked the year before. His name went onto the list of prospective fliers along with people who would later make their names flying on the shuttle, like Karol Bobko, Henry Hartsfield, Gordon Fullerton, and future NASA administrator Richard Truly. The MOL program took the rookie astronauts to Vandenberg Air Force Base, where the spy stations were meant to launch into an orbit circling the North and South Poles, as opposed to the more familiar path around the equator taken by Apollo and the yet-to-be-built space shuttle.

Crippen recalled how the MOL program got to the point of real hardware and simulators in which the two member crews would train. "It looked pretty much like what would go into Skylab," says Crippen. "We had eating facilities, a latrine, and sleeping accommodations, which were basically sleeping bags on the walls." The later Skylab space station program in the early 1970s would give crews a lot more interior space to float around in, but that didn't matter to Crippen and his MOL colleagues. They were close to flying.

There was one nagging worry about the vehicle, however. It was the hatch.

Once in orbit, the MOL astronauts would open a door built into the heat shield on the blunt bottom of their Gemini craft and float inside the lab for a mission designed to last about a month.[8] Cutting holes in a heat shield would later be commonplace on the shuttle, where doors had to be included on the belly for the landing gear to pop out. The wheels would drop down after the dangerous reentry into the earth's atmosphere, where the spacecraft faced 3,000 degrees of heat due to friction with the air. The shield had to work or the vehicle would burn up. The shuttle and its modified heat shield were built in the 1970s, though, and MOL flew in the mid-1960s. Back then, cutting holes in the heat shield of a space capsule was considered pushing the envelope. "That was one of the big concerns," says Crippen. "We were going to launch a test vehicle to prove it [the hatch] had integrity."

As it turned out, the MOL astronauts had more to deal with than the experimental heat shield. There came a date Crippen remembers by heart: "June tenth, 1969. It's one of those days that are burned into my memory."

About a month before Neil Armstrong set foot on the moon, the Pentagon abruptly cancelled the Manned Orbiting Laboratory. That left Crippen and his colleagues stuck on the ground without a single flight. The space veteran remembers this as the low point in his life. Still stinging from the news, Crippen and his former MOL colleagues were gathered at their customary Monday morning meeting when someone floated a radical idea. "Well, we were sitting around wondering what to do now," Crippen says, "and Karol Bobko piped up and said, 'Why don't we call NASA?' We all pooh-poohed the idea, but then we made the call, and within a week or so we were talking with Deke Slayton." Slayton was the chief of the Astronaut Office, and their last chance to stay somewhere in the space program.

The MOL astronauts didn't harbor much hope, since NASA was already starting to cut back on moon missions and there were plenty of seasoned candidates in jeopardy of losing their own seats to the lunar surface. Still, Crippen and a number of the other grounded MOL candidates were chosen to join NASA and perform supporting roles in missions like the final moon flights, the Skylab orbital workshop, and the Apollo-Soyuz docking trip conducted by the United States and the Soviet Union. "Even when you think you know the answer," says Crippen about his prospects with NASA, "it never hurts to ask."

It took fourteen years for Crippen to finally get a flight assignment on the first space shuttle mission. He admits he thought several times about quitting, but each time decided to stick it out. NASA thought enough of the "MOL men" to send at least one of them on each of the first eight shuttle flights.[9]

When the demise of the Apollo program left a number of longtime astronaut candidates still waiting for their chance at a space mission, their only choices were to quit or get in line with the MOL astronauts and wait for the shuttle to go. Even before Apollo closed down, NASA was looking at trimming back moon missions. Money was drying up from Washington, and the casualties were Apollo flights 18, 19, and 20. No official crew assignments had been made for these missions, but the pattern of astronaut rotation that moved people from backup to primary crew members gave a strong indication of who lost their chance. *Apollo 12* command module pilot Richard Gordon orbited the moon on that flight while Pete Conrad and Alan Bean walked on the Ocean of Storms. Gordon was apparently due to lead *Apollo 18* and make a landing of his own. That flight never happened, and the closest Gordon would come to being on the moon is a painting by crewman-turned-artist Al Bean depicting all three *Apollo 12* astronauts on the lunar surface. Gordon is in the middle, with Bean making "bunny ears" with his gloved fingers behind Gordon's helmet.[10]

Now, with the impending arrival of the CEV, it's the turn of NASA's current corps of space shuttle astronauts to feel like they're on the endangered species list. Even being an astronaut in the shuttle program isn't a ticket to instant stardom, let alone long-term job security.

Take astronaut Carl Walz, for example. He was the man who chaired one of the panel discussions during NASA's 1st Space Exploration Conference at Walt Disney World in 2005, and who made the comment on NASA's lack of technical know-how for a manned Mars mission that led off this book. His face may not elicit immediate recognition, but his lapel pin just might. It's a small gold star going through a tiny hoop trailing three beams of light. Mercury astronauts Gus Grissom and Gordon Cooper designed the pin, and you have to go to space to get one. Astronaut candidates get a silver version of the pin to wear before their first flight. Walz earned his gold pin in 1993 with a ten-day mission aboard space shuttle *Discovery* that launched the Advanced Communication Technology Satellite.[11] The mission also gave people on Earth close-up pictures of the shuttle's tile-covered hull, taken by the Orpheus-SPAS satellite built by Germany and deployed by the astronauts. Walz topped off his rookie flight by doing a spacewalk to test tools that would later be used to service the $1.6 billion Hubble Space Telescope.[12] There's an old saying that goes "If you don't know jewels, know your jeweler." In Carl Walz's case, that specialty is space travel.

Few people approached Walz after the group adjourned during the Orlando space conference. No autograph seekers, no one asking to have their picture taken with him.

Being an astronaut is different now. At one time, flying in space for NASA meant getting your picture on the cover of *Life* magazine. Now the

Figure 8. Shuttle astronaut Carl Walz giving a thumbs-up during a fit check of his launch-and-entry suit prior to liftoff of STS-65, NASA's 2nd International Microgravity Laboratory mission. Courtesy of NASA.

job of an astronaut is described as a kind of fame you "put on and take off." Someone in the Astronaut Office coined that phrase to explain the impact of the blue flight coveralls that shuttle veterans wear when they want to attract attention. Walz's flight suit is festooned with embroidered NASA flight crew patches stitched on the pockets and sleeves. Each is a trophy of a successful mission in orbit. One patch is the so-called NASA "meatball." That's the agency's round blue patch with the red slash through the middle dating back to the early Mercury days. The astronauts design a new crew patch to symbolize each flight they make. Walz's first one was from STS-51, the 1993 mission with the communications satellite and his walk in space. Its circular patch is mostly black with an image of the shuttle soaring into orbit. The communications satellite is shown flying off to the left in the form of a yellow star, and the second satellite, the one that photographed the shuttle, is depicted with the three stripes of the flag of Germany. Walz's accomplishments are apparent when he's in his blue flight coveralls. When he's dressed like a civilian, you might be standing next to him and not know it, but the garb of an astronaut brings instant recognition that this person has "been there."

Canadian astronaut Marc Garneau once demonstrated how much weight NASA gives to the impact of a blue flight suit. When he showed up at the Kennedy Space Center for some press interviews, the first questions were from radio reporters, and it was clear no one in the listening audience could see him dressed in slacks and a shirt and tie. But then one of his handlers said CNN wanted to talk with him on camera. That impelled Garneau to rush with his duffel into a vacant office at the KSC press center and make a quick change into his blue coveralls. Being the first Canadian in space might draw crowds at the ice cream stand in Ottawa, but elsewhere Garneau was just another face in the crowd.[13]

NASA says the hurdles that Carl Walz, Marc Garneau, and Bob Crippen had to clear to qualify for the elite astronaut corps will not change much with the arrival of the new Crew Exploration Vehicle. There will be extensive physical and mental testing, and only the most qualified people will be chosen to train for a mission. In the 1950s and 1960s, test pilots took center stage during the early Mercury and Gemini flights that preceded the Apollo lunar landings. On Apollo, NASA expanded its horizons to include people with engineering and science backgrounds. Many of the mission specialists, who occupied nonpiloting seats on the space shuttle, held advanced degrees in science and technology. Trips to Mars may require certain specialties like geology or a medical background.[14] But even an impressive academic record is no guarantee of flying into space.

Veteran astronauts Rick Husband and Kalpana Chawla were both members of NASA's astronaut class of 1995, and later died in the *Columbia* accident in 2003. The space agency picks future crew members in groups, or classes, to undergo basic training and then wait for a mission assignment. One face not included with Husband's and Chawla's in the class portrait for 1995 was that of Dr. Victoria Coverstone, a professor of aeronautical engineering at the University of Illinois at Urbana-Champaign and a pro-spective astronaut candidate that year. "Ever since I was seven, I wanted to be an astronaut," said Coverstone. "When I was little, I used to read these Nancy Drew mysteries, and I always liked finding the answers to things I didn't know about." Coverstone had already worked with the Jet Propul-sion Laboratory in Pasadena on new technologies. One idea was a solar sail that could propel small unmanned spacecraft by catching a stream of particles from the sun, the way a sailboat catches the wind.

NASA wanted Coverstone to apply to be a mission specialist. That's the type of shuttle astronaut who works the robot arm and does space-walks but doesn't pilot the space plane. The process started with a stack of standard government employment forms. Coverstone applied in 1994, before the Internet caught on. She submitted the paperwork by mail, lick-ing stamps to send the completed pages back.

"I didn't hear from NASA for weeks," said Coverstone. "But then my references called to say that they'd been questioned by the government." Later the phone rang.

Coverstone was asked to go to Houston for interviews and tests. She and twenty other hopefuls crowded into a conference room at the Johnson Space Center. Among those applying to be an astronaut was Rick Hus-band, who would later die aboard shuttle *Columbia*. Coverstone's group filled out personality profiles and then were interviewed by a panel led by Apollo astronaut John Young, who commanded NASA's first space shuttle mission. "He asked us to talk about what we did in high school," said Cov-erstone with disbelief. "Here was a man who walked on the moon. What could I have done in high school that would possibly be of interest to him?"

After the question-and-answer session, all the candidates were loaded into vans and driven to a building with no windows. There they would sit alone in smaller rooms and take more exams, which included an essay question. Coverstone was asked to write why she wanted to be an astro-naut. "I wrote mine like a patriotic red-white-and-blue sort of thing," said Coverstone. "One woman in our class wrote hers like a David Letterman 'top ten.' Number ten was getting all those frequent flier miles, and she went on from there."

Neither of them made it.

"It felt like being sucker punched," said Coverstone. "But I kind of felt I wasn't going to pass that eye exam." Being a robotic arm operator on the space shuttle requires a sharpness of visual depth perception that flight surgeons decided Coverstone lacked. One school of thought among successful applicants is that NASA respects people who don't give up. Ironically, candidate William McCool, like Victoria Coverstone, was rejected this time around. He applied again in 1998, was accepted, and then joined Husband aboard *Columbia*'s ill-fated flight in 2003.

Even having sat next to Rick Husband and McCool and, in hindsight, knowing their fate aboard *Columbia* doesn't dampen Coverstone's enthusiasm for the program and her sense of loss for not getting in. "Astronaut Steve Robinson was part of my application group, and he got in," she said. "After *Columbia*, he was named to the first mission to go following the accident. If I had made it, I would have been perfectly positioned at the right time." Now, that right time may fall to the kids sitting in Coverstone's aeronautical engineering classes. Many of these youngsters considered the space shuttle boring and the astronauts who fly them "old fogies."[15]

Back during the earliest flights of the space shuttle, starting in 1981, NASA's choice of crew members was decidedly fogies-first. Before scientists and technicians got to fly, the hardened test pilots would make the first four "experimental flights" of the new winged spacecraft.

The lifetime of the shuttle has been marked by a lack of focus and a series of compromises, which fueled criticism of the program from the very beginning. In the late 1970s that didn't matter to men like Manned Orbiting Laboratory crew member Bob Crippen, who was training hard for his first flights in orbit. Back then, everything about the space shuttle was shiny and new. Crippen was originally selected to fly in space capsules. Now he would strap himself aboard a spacecraft with wings.

3

1981

The Path to STS-1

It would be a simple matter to think of NASA's first space shuttle, poised for its maiden liftoff from the Kennedy Space Center in 1981, as the "start" of the shuttle program. Even the date of the first launch of *Columbia* gave NASA the opportunity to upstage the robust Soviet space program, which had been regularly launching crews to small space stations while NASA sat on the ground after the end of Apollo in 1975. Shuttle *Columbia* was set to blast off on the twentieth anniversary of Russian cosmonaut Yuri Gagarin's departure as the first man in space in 1961. "Take that, Moscow," the flight seemed to say as America roared back into the space game.

But by the time *Columbia* was bolted to its now familiar external fuel tank and with the twin gleaming solid rocket boosters, NASA was at the end of a decadelong struggle to keep its astronauts flying following Apollo. Unlike the shuttle, the moon program had a mission, but not one that ensured long-term success. Even as the agency took its bows for Armstrong's "one small step" in 1969, a contentious debate was under way on what the

United States' next step into space should be. The budget was cut back, and NASA was left to scramble for a way to keep Americans in space. The agency would have to be satisfied with a space plane built to work no higher off the ground than low earth orbit, and perform tasks within the agency's reduced grasp.[1] NASA first sold the space shuttle as a way to make space travel routine, accessible, and cheap in the face of modest post-Apollo budgets. The original plan called for a two-stage fully reusable vehicle.

The first part was to be like a winged fuel tank with a cockpit inside. Two astronauts would pilot that craft, with a second winged section, called the orbiter, carried piggyback. The first stage would ignite and carry itself and the orbiter off the launchpad. Once the tandem vehicle made it to the edge of space, the shuttle section would detach and fly the rest of the way on its own. The first-stage vehicle would be piloted back to a landing at the Kennedy Space Center.

Ten billion dollars to build, NASA said.

Too expensive, Washington replied.

So was the plan to build a manned orbiting space station for the shuttle to visit. NASA might get to start one of its new projects, but not both at the same time. The agency would have to settle for the shuttle, and for half the money it was seeking. That spurred some political wheeling and dealing where it seemed nobody got what they really wanted.[2]

Since the big reusable shuttle was unlikely to fly in Congress, NASA toyed with the idea of creating a much smaller shuttle with a wedge-shaped design. TV viewers in the 1970s saw this type of spacecraft during the opening credits of the show *The Six Million Dollar Man*, where a fictional crash injured the character Steve Austin, who then was made into the bionic man who ran in slow motion to save the world every week. The accident footage used at the start of each episode was of an actual NASA test flight that went bad. The Northrop Corporation built a vehicle called the M2-F2 as a possible prototype for a NASA shuttle.[3] The vehicle did a belly flop on the runway in 1967 and tumbled six times before coming to rest. The test pilot was seriously injured, but not fitted with bionic replacement parts. The smaller flying-wedge concept would later make another splash in popular culture in the 1980s as the rebel Snow Speeder craft in the second *Star Wars* film, *The Empire Strikes Back*.

The smaller vehicle may have suited NASA and Hollywood. But it didn't sit well with the people the space program needed on its side if the shuttle was going to be funded. NASA recognized it needed an ace up its sleeve to win the support of skeptical lawmakers. The answer was more compromises and a plan to involve the Pentagon. In the end, a bigger space plane would survive the process, not a smaller one.

The U.S. Air Force's interest in manned spaceflight had cooled significantly since the days when the X-15 rocket plane took pilots to the very rim of space. A later version, the rocket-propelled space plane called DynaSoar, had been shelved in 1963. The Manned Orbiting Laboratory space station met a similar fate in 1969. The air force appeared content to confine its space-related ambitions to spy satellites launched on unmanned rockets.[4] That is, until some creative arm-twisting on the part of NASA. The agency proposed using its new space shuttle as a cheap and reliable way for the air force to deliver payloads to low earth orbit. NASA's strategy was to have a willing customer, ready to go, when it asked Congress for shuttle funding. The spacecraft would be sold as a way to make money, not just spend it, as 30 percent of NASA's shuttle missions would include air force payloads. In the process, the air force would move to retire its fleet of Titan rockets, which had been the military's workhorse launcher up to that point.[5]

The plan seemed to cover all the political bases, but that was little comfort for some people who sat in on the negotiations. "This was kind of shocking," says former U.S. secretary of the air force Robert C. Seamans Jr. "Why put astronauts at risk to put unmanned satellites in space?" Seamans's concerns came from his unusual vantage point, which allowed him to see both sides of the NASA–air force collaboration. From 1965 to 1968 as deputy administrator of NASA he had overseen the launches of the two-man Gemini missions that sharpened many of the skills the astronauts would need to successfully land on the moon. In 1970 President Nixon had chosen Seamans to be secretary of the air force. Then NASA came knocking at the door to make the shuttle the military's only ticket into space. "I thought it was asinine," says Seamans, "to launch every payload on the shuttle, risking three to five astronauts per flight."

Still, advisors to the air force thought the idea was worth examining. NASA appeared eager enough to tailor the shuttle to meet military requirements, and the air force had a list of things it wanted to see on the new spacecraft.[6]

The Air Force Gives NASA Its "Laundry List"

The military had spy satellites the size of railroad boxcars to send into space. If NASA was going to carry those cargoes, a bigger spacecraft was needed. That sent NASA back to the drawing board to design a winged shuttle with a cavernous cargo bay, right behind the crew cabin where the astronauts would live and work. Critics would later complain that the massive compartment would fly mostly empty for much of the shuttle's lifetime.

That wasn't the end of the air force's demands. It also wanted to launch the shuttle from Vandenberg Air Force Base in California instead of from the Kennedy Space Center in Florida. Launches at Kennedy traditionally sent spacecraft on a flat path around the equator like a vinyl record spinning on a turntable. California launches would mean an orbit around the earth's poles. This fit the Pentagon's needs, since military planners wanted the shuttle to make one trip around the planet and then land immediately. That would give the astronauts time to deploy their spy satellite without sailing over hostile territory during open warfare.

It also created a design problem for NASA. During such a quick space trip between the North and South poles, the globe would still be rotating on its axis. So, as the Orbiter flew around the Earth from North to South, or South to North, the vehicle's landing spot in California would be moving in a direction of its own, to the East. If the future shuttle dropped straight from orbit on its path around the globe, instead of landing on dry ground, it would hit a patch of the Pacific Ocean off the California coast. To make safe touchdowns possible, the solution was to give the shuttle the ability to glide off its orbital path, or "downrange" as NASA likes to say. So, to sum it all up, the need to glide to the ground is why the shuttle had delta shaped wings. The need to carry big spy satellites drove NASA to give the craft its cargo bay.[7]

The case for shuttle funding didn't save NASA from critical compromises on the spacecraft's design. The amount of money Congress wanted to spend and the demands of the air force required designers to outfit the shuttle with reusable twin solid fuel boosters that would provide about 70% of the thrust needed to send the shuttle to orbit. The boosters would be bolted on either side of a completely disposable external fuel tank with the orbiter riding bolted to the side. The tank would feed cryogenic

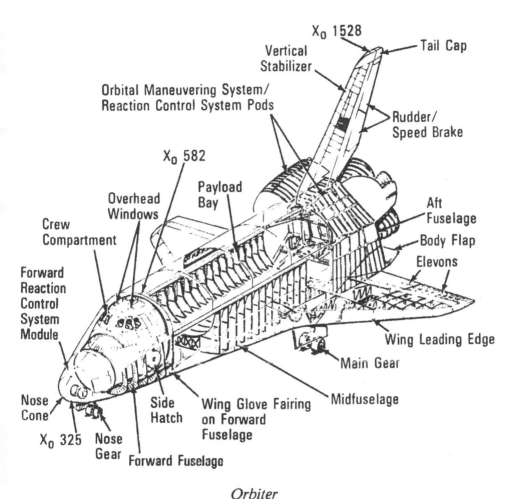

Orbiter

Figure 9. Structure of the Space Transportation System, or shuttle. Roughly the size of a DC-9 aircraft, the orbiter without its heat tiles resembles a lime green airplane. The pressurized two-story crew cabin at the front is where the astronauts live and work. The roomy payload bay can accommodate cargoes like satellites, a European-built Spacelab module for experiments, or smaller Spacehab compartments for either science work or storage of supplies for the Russian space station Mir or the International Space Station. Courtesy of NASA.

liquid hydrogen and oxygen propellant to three main engines at the rear of the shuttle beneath the tail rudder. Once the boosters dropped away two minutes after liftoff, the shuttle's engines would provide thrust for the remainder of the trip into orbit. For Robert Seamans, the changes to the spacecraft made a shaky deal even worse. "It seemed like a good idea," he says about the partial re-usability of the shuttle. "But, it was like throwing away a 747 after every trip across the Atlantic."

Like it or not, the deal was done and the die had been cast. NASA would have to make the best of it.[8]

NASA Got Its Shuttle—Now It Gets Ready to Launch It

While NASA haggled with Congress and the air force over the design of its new shuttle, engineers were already hard at work on the new vehicle. For one man, a shuttle-related job began while the astronauts on the first manned moon landing were still trudging around the dusty lunar landscape.

In mid-1969, NASA engineer Sam Beddingfield just wanted some time off. A long-awaited vacation was particularly welcome now, since his worst nightmare about the first manned moon landing had been dispelled. "When we sent the lunar module down to the moon's surface, we didn't know if it would sink or not," Beddingfield says. It didn't, and the lunar lander Eagle was on solid ground on the moon. Having weathered that case of butterflies in the stomach, Beddingfield headed home to North Carolina. The break didn't last long: a phone call was waiting for him from NASA. They wanted him in Alabama for a new project. It would be called the space shuttle.

"NASA wasn't too sure how it was going to work," Beddingfield recalls. For him, it was a familiar refrain. He had been dealing for ten years with an agency that wasn't certain how to make its projects fly.

Before joining NASA, Beddingfield had spent part of his career flying jets for the U.S. Air Force. That included work at Edwards Air Force Base, where many of America's top test pilots gained experience flying high-performance aircraft. Chuck Yeager broke the sound barrier at Edwards in 1947, but it was Beddingfield's close friend Gus Grissom who earned international acclaim as one of the original Mercury Seven. Astronaut Grissom would go on to make a controversial suborbital flight from Cape

Canaveral to the Bahamas aboard the Mercury capsule dubbed Liberty Bell 7. After splashdown, the main hatch blew open prematurely and the craft sank. Many blamed Grissom for the mishap, and Beddingfield spent decades defending his friend from critics.

Back in 1959, before NASA and Project Apollo, Beddingfield left the air force to pursue farming on land owned by his father in North Carolina. There was only one catch. "After three days," says Beddingfield, "I figured out I didn't give a damn about farming."

With a wife and family to provide for, he went back to the only thing he knew, aviation. Beddingfield's search for a job led him north to Langley Field in Virginia. The people in the employment office said they had no aviation jobs available, but the news wasn't all bad. They said there was a new outfit called NASA on the other side of the runway that needed people. Why not try there? Beddingfield drove around to the NASA office, walked inside, and behind the desk sat Gus Grissom.

"What are you doing here?" the fledgling astronaut asked.

Beddingfield said he was looking for a job, but Langley didn't have any.

"Well, we have jobs here," Grissom said.

Beddingfield admitted he didn't know the first thing about rockets.

"That's okay," Grissom responded. "Neither does NASA."

And that's how Sam Beddingfield found himself a key engineer on the project called Mercury that sent men like Alan Shepard and John Glenn into space. His buddy Gus Grissom would team up with rookie John Young on the first flight of the two-man Gemini capsule. Sam worked on that, as well.

As Beddingfield became the first Kennedy Space Center engineer to switch from Apollo to the shuttle program, NASA was still smarting over the scuttling of its plans for a fully reusable spacecraft. The craft NASA got, with its boosters and disposable fuel tank, left engineers like Beddingfield scratching their heads.

To NASA personnel who were used to launching capsules on rockets with the pointy end up and the engines aimed straight down, the finished spaceship was funny looking. The now familiar shuttle seemed to dangle from the side of the 154-foot-tall external fuel tank. The first four tanks were painted white, until NASA realized the paint weighed about seven hundred pounds. The agency decided it was better to devote that weight

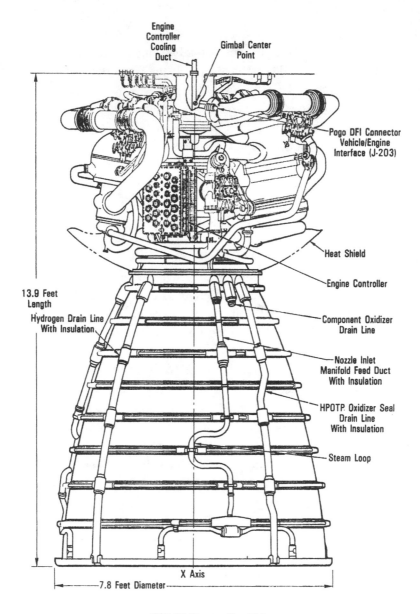

Engine
Controller
Cooling
Duct

Gimbal Center
Point

Pogo DFI Connector
Vehicle/Engine
Interface (J-203)

Heat Shield

Engine Controller

13.9 Feet
Length

Hydrogen Drain Line
With Insulation

Component Oxidizer
Drain Line

Nozzle Inlet
Manifold Feed Duct
With Insulation

HPOTP Oxidizer Seal
Drain Line
With Insulation

Steam Loop

X Axis

7.8 Feet Diameter

SSME Controller Side

Figure 10. Diagram of a space shuttle main engine or SSME. The three main engines on the spacecraft supply it with 20 percent of the thrust needed to put it into orbit. After the solid rocket boosters exhaust their fuel and are jettisoned, the main engines continue operating for six minutes or so. The white smoke belched from the engine nozzle during ignition and flight is steam. Courtesy of NASA.

to cargo instead of paint, so the tank's orange-colored foam insulation was left plain. The space shuttle also behaved in a way unfamiliar to veterans like Beddingfield. Instead of leaping off the pad upon engine ignition, the orbiter's three liquid-fueled main engines expelled white steam at a slight angle with the craft still held down. "The whole thing leaned forward," says Beddingfield. "Then, when it came back up, the solids [booster rockets] would fire and then you'd go." Shuttle crews would note that this leaning effect, known as the "twang," didn't go away after liftoff. The spacecraft would continue to "shimmy" back and forth as it made the trip up toward orbit.[9]

Odd, maybe, but at least it received funding from Washington.

Columbia Makes Its Debut

"I was just lucky they picked me for it," says astronaut John Young. NASA knew early on which astronauts it wanted to fly the first experimental mission of the untested space shuttle. In 1981, Young was already a veteran of two Gemini flights and two missions to the moon. That included landing the lunar module Orion on the Descartes highland plains for a moonwalk of his own. "I was the only one in the office with all that experience," says Young, "but there were a bunch of Apollo guys who wanted it." Besides Young, there was a small remnant of the original astronaut corps including his crewmate Ken Mattingly from *Apollo 16*, along with Jack Lousma and Paul Weitz, who both flew on Skylab, and Vance Brand from Apollo-Soyuz. They were in line along with the grounded crew members of the air force's Manned Orbiting Laboratory, like Bob Crippen. The process of picking the first two men to fly space shuttle *Columbia* was a crash course for some, a refresher for others, in the mysterious way NASA selected crews for space missions. "I thought they would team up two veteran astronauts for the first flight," says Crippen. "No one was more surprised than me when they picked me." The news came following one of the approach-and-landing tests featuring the shuttle vehicle called *Enterprise*. That craft never made it into space, but it confirmed that the shuttle's design was sound enough to glide to the ground and make a controlled touchdown. The craft was carried up by a NASA 747 jet and then cut loose to land on its own. Crippen was inspecting *Enterprise* with George Abbey, director of the Johnson Space Center, after the test. "Abbey just said,

'Well, Crip, how would you like to be on the first one?'" Crippen says. "I was doing handsprings for the rest of the day." NASA had its crew for the first launch of the shuttle.[10]

As the process of building *Columbia* went from the drawing board to nuts and bolts and cutting metal, John Young and Bob Crippen were deeply involved in the space plane's development. Crippen led a group of six hundred technicians writing software for the vehicle's five onboard computers. As the shuttle's first commander, astronaut Young spent most of his time haggling with NASA over the cockpit.

One battle was over the "heads-up display." That's a clear plate of glass on the dashboard of an aircraft, so the pilot can look straight ahead through the windshield while numbers like airspeed and the craft's altitude and attitude are projected onto the glass, between the pilot and the windshield for him to read. "I really had to sell the heads-up," Young says. "Deke [Slayton, chief of the Astronaut Office] said, 'You don't need that. Just fly by the seat of your pants.' Well, you don't want to do that on the shuttle. You better not do that." The need for a heads-up display was reinforced by the speed at which the orbiter would be traveling during landing and the steep angle it would take. Even though it was advertised as a glider, Young compares flying the shuttle to flying a brick. The landing is also referred to as a controlled plummet. The spacecraft would dive toward the earth at an angle six times steeper than a commercial passenger jet. Some veteran crew members compare it to a dive bomber run.[11]

Even seasoned shuttle pilots describe their first landing in the spacecraft as the scariest part of their rookie missions. That includes NASA's first female shuttle pilot, Eileen Collins, who went on to become the agency's first female shuttle commander. She also led the first mission to fly after the *Columbia* disaster. Her first touchdown was during her rookie mission, which included the first rendezvous between a NASA orbiter and the Russian space station Mir. "It was certainly stressful," said Collins. "As you come down and gravity comes back, you feel your helmet squashing down on you and the heaviness of your suit, and if certain items aren't positioned right, they start to fall around the cabin." Following this experience, Collins made it a point to train her rookie pilots to handle that part of the flight.[12]

Though he was active in designing the space shuttle, Young takes no credit for the layout of the cockpit where he and Crippen would sit

Figure 11. The space shuttle's updated flight-deck dashboard, known as the glass cockpit. The brightly colored computer screen replaced the 1970s-era mechanical instruments sardonically called "steam gauges." More energy-efficient, the new technology allows the astronauts to use computer screens previously switched off to save power. Courtesy of NASA.

strapped in their flight couches during the trip to orbit and the dangerous reentry into the earth's atmosphere. Later in the program, members of the astronaut corps would grouse over the mechanical dials, switches, and gauges on the orbiter's dashboard that dated from the 1970s. The clunky design was referred to in the aviation community as "steam gauges," and the equipment on the vehicle would remain a bone of contention for years. Military fighter jets and commercial passenger planes had long since moved on to brightly colored computer screens called "glass cockpits" while the shuttle chugged along with its older mechanical technology. It was a great irony that the most sophisticated aircraft in the world had a dashboard dating back to when disco was king.

Astronaut pilots like Scott "Doc" Horowitz admit being teased about the shuttle's quaint dashboard by fellow aviators whose aircraft had glass cockpits. "It's okay," Horowitz says. "I got 'em beat when it comes to thrust-to-weight ratio." Translated, Horowitz means he goes fast enough to achieve orbit and his buddies don't. Adding to the indignity of outdated equipment, launch managers ordered shuttle commanders not to switch on some of the CRTs or computer screens in the older steam-gauge cockpits in order to save electricity during liftoff. This prevented a third crew member on the flight deck, acting as flight engineer, from monitoring key systems. That duty fell back on commanders and pilots, most of whom felt they had enough to do during ascent and landing.[13]

The instrumentation wasn't the only source of concern on the fledgling shuttle. Its solid rocket boosters were another. Up to then, solid fuel had never been used on launches with people aboard, because you can't shut the boosters down after ignition. Engineers are blunt on the matter. They say if the astronauts ran into trouble, no emergency safety aborts would be possible until after the solids ran out of fuel and were jettisoned. That left Young deeply concerned about the trip to orbit, since he recalled only about five test firings of the new boosters, with each lying on its side. The first time the boosters would be used for launch would be with the two astronauts strapped inside the shuttle. There was no plan to man-rate the space shuttle before putting people aboard.[14]

Man-rating is a process where unmanned versions of a spacecraft are launched to make sure everything works, to help guarantee the safety of the crew. The challenge in sending up the shuttle unmanned was making a successful landing. *Columbia*, the first orbiter, didn't have an autopilot system. A combination of budgetary constraints and emotions kept NASA from giving the shuttle the ability to fly itself. "The contractors said it would force us to slip [delay the maiden flight] by six months and cost about one hundred million dollars." Young recalls. "It really would have been a half billion dollars, because there were seven hundred and fifty-four switches in the cockpit that would have to go through the software." There was also resistance from at least one member of top NASA brass, Young says. John Yardley was in charge of developing the new vehicle. This job followed a long career at NASA where he helped design the Mercury capsules that carried men like John Glenn and Gordon Cooper to orbit. Yardley was firmly against the idea of sending the space shuttle up unmanned on its first flight. "John Yardley said he didn't want his airplane flying unmanned over California," says Young. "He didn't want to do that. 'Cause there were a lot of people out there he cared about." There were also explosive ejector seats on the early test flights, but that didn't provide much comfort for either Young or Crippen. Their years spent flying jets for the U.S. Navy had taught them one thing. "The first time you use an ejector seat on a new aircraft, it usually doesn't work out too well," says Young. "So I wasn't looking forward to using it on the shuttle." Also, NASA realized that if the crew ejected from the spacecraft with the solid rocket boosters still attached and burning, the astronauts would fall back into the thrust coming off the rocket nozzles and be incinerated. The escape seats and their explosive charges to blast them free would be removed when the shuttle was declared operational after the first test flights.

The first crew of the shuttle was, if not resigned, prepared to test the space plane while sitting in the cockpit. "The shuttle was designed to fly with people in it," says Bob Crippen. "And we wanted to do what it took to ensure mission success, so we were ready to do that."

One additional point of concern was the heat protection tiles. Mercury, Gemini, and Apollo capsules had heat-resistant material on the blunt end of the craft for a one-time-only plunge through the earth's atmosphere at the mission's end. Each of the orbiters planned for the shuttle fleet was meant to make one hundred flights or more, so NASA wanted something more permanent. The light ceramic tiles, later to be augmented with protective cloth blankets, seemed like the right idea. The shuttle, with the tiles peeled off, looks like a lime green airplane. Its aluminum skin can take only about 350 degrees of heat before it melts. Something would have to insulate the craft and the crew from the 3,000 degrees of heat generated as the shuttle's belly scraped against the atmosphere during reentry. The tile idea also led to regular meetings that drove John Young crazy. The engineers, he says, "kept on holding these meetings every month for two years talking about how the tiles would fall off and burn holes in the shuttle and stuff. After the flight, I looked around and no tiles came off the bottom."

Prior to the maiden launch of *Columbia*, Young was already NASA's most experienced space veteran. He was known for being soft-spoken but no-nonsense. Young nettled NASA managers by smuggling a corned beef sandwich aboard the capsule for the maiden voyage of the two-man Gemini program. Not even his crewmate, Gus Grissom, knew about it.

The first launch of the space shuttle contained a lot of unknowns, and that left engineer Sam Beddingfield worried. "We didn't know would it launch right," he says. "And would the [heat] tiles work during reentry?"

Buckle Your Seat Belt and Let's See If Columbia Works

On April 12, 1981, the countdown clock was slowly ticking to zero. *Columbia* was a major departure from the spacecraft that Young flew during Projects Gemini and Apollo. Climbing in is just one example. The Gemini capsule had two hatches, one over each of the flight couches. Pad crews opened the hatches, and each astronaut slid in and was strapped into the seat. The Apollo capsules had one hatch within inches of the three seats for the crewmen.

Columbia was different.

On launch day, the astronauts would ride to the launchpad in a small silver bus called the AstroVan. They would travel up the gantry in an elevator and then walk across a gangway, called an access arm, to the waiting spacecraft. At the end of the gangway is a compartment called the white room, which nestles up against the vehicle. Straight ahead, there's a round hatch on the port side of the spacecraft that leads into the lower half of the crew compartment. You can't walk in. Rather, astronauts get down on their hands and knees and crawl in.

The view inside is a little topsy-turvy, since space shuttles launch pointing up but are built to land like an aircraft. That means as you go into vehicle, the "wall" to your left on launch day is the "floor" on landing day. The "wall" on the right is the ceiling of the lower level of the crew cabin. The crew members crawl in and then make a sharp right-hand turn and squeeze into the "upper" level of the cabin, or cockpit. For STS-1, there were just two seats, for the commander and the pilot. When extra crew, called mission specialists and payload specialists, were added later, the view inside the vehicle was even stranger. There would be two extra seats installed directly behind the commander's and pilot's. On the lower level, or middeck, additional flight seats for the fifth, sixth, and seventh members of the crew would appear to be bolted to the "wall" on the left side.[15] On launch day, the view from the flight deck was not the unobstructed

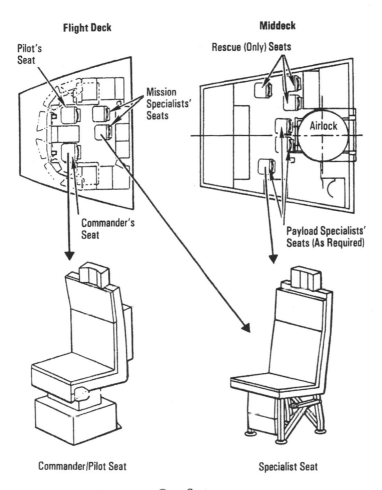

Flight Deck

Pilot's Seat

Mission Specialists' Seats

Commander's Seat

Middeck

Rescue (Only) Seats

Airlock

Payload Specialists' Seats (As Required)

Commander/Pilot Seat

Specialist Seat

Crew Seats

Figure 12. Seating diagram of a space shuttle's upper level, or flight deck, and lower level, or middeck. Under normal gravity the astronauts climb from one deck to the other on a ladder. While weightless in orbit, they float through one of two openings. The middeck also features an airlock to permit spacewalks outside the shuttle. Courtesy of NASA.

skyscape that John Young knew from previous missions. Gemini and Apollo capsules were on the tip-top of the rockets. The shuttle was attached to the side of the external fuel tank, so the tank and the nose cones of the two solid-fuel boosters were visible through the windows.[16]

The crew would also be dressed differently for a shuttle launch than for Young's most recent blastoff on *Apollo 16*. In 1972 Young, accompanied by crewmates Charlie Duke and Ken Mattingly, would make NASA's next-to-last manned moon landing. The three astronauts wore bulky white spacesuits that were standard for Apollo, giving them the appearance of the Michelin Man or perhaps the Pillsbury Doughboy. For the maiden launch of space shuttle *Columbia* in 1981, Young and Crippen would wear tan-colored pressure suits that made them look like futuristic fighter pilots. Their equipment would include survival gear in case they needed to eject from the craft during a midair emergency. Instead of the fishbowl-style helmets worn during Apollo, the first shuttle crew used white helmets with wide clear visors.

Once suited up, the *Columbia* astronauts were ready to head to the pad. "It was all pretty routine." That's how John Young recalls the maiden launch of the space shuttle.

The "New Guys" Step Up

The blastoff of STS-1 was far from routine for the members of the astronaut class of 1978. They were the first group of rookies NASA had hired to fly in space since 1966. The list of applicants included names that would never become household names like Neil Armstrong and John Glenn, but many did become synonymous with the space shuttle. One of them, Robert "Hoot" Gibson, flew the first docking mission between the shuttle and the Russian space station Mir. Norman Thagard became the first American to ride a Russian Soyuz rocket and live and work for an extended period on Mir. Sally Ride was the first American woman in space, Kathy Sullivan made the first U.S. spacewalk by a woman, and Guy Bluford was the first African American in space.

Another name in the astronaut class of 1978 was Dan Brandenstein. When Young and Crippen blasted off on the first launch of the shuttle, Brandenstein would be their sole link to the ground. The job is called capsule communicator, or capcom. "I sort of inherited it," says the four-time shuttle veteran. "I was backup capcom to [Skylab astronaut] Ed Gibson. Then he retired from NASA, and I got it." Brandenstein would pilot the first night launch and night landing of the space shuttle two years after STS-1. But he recalls being capcom as a great job. "It's the next best thing to being on the shuttle," he says. "When you're on a mission, you don't get to enjoy the thrill of the ride because you're focused on the job of flying."[17]

When Dan Brandenstein was assigned to be capcom on the first launch of the space shuttle *Columbia*, there was no room for chatter between Mission Control and the flight crew. Everything Young and Crippen would need to check and relay to Houston during liftoff was written on cue cards tacked everywhere inside the cockpit. Brandenstein wrote most of them with the help of his backup, astronaut Terry Hart. "We were on 'cue card alley' back then," says Brandenstein. "We even had neckties made up with all of the cue card phrases on them, stenciled upside down. All you had to do was look down to read them. We gave one to the NASA administrator and one to Young and Crippen." The ties weren't meant as a fashion statement, but rather as a way to break the tension on a maiden shuttle flight with a lot of tension to go around. "To go from the launchpad and eight minutes later to be going seventeen thousand miles per hour," says Crippen, "I can't equate that ride to anything else."

Young's assessment of the ride was different, for it was based on riding Gemini-Titan rockets and two Apollo Saturn rockets on his previous flights. His attention, like Crippen's, was focused on the abort modes, in case something happened following ignition of the three main engines and the two solid rocket boosters. There was talk of making the mission of STS-1 an abort test, with Young and Crippen trying to bring *Columbia* back to the Florida landing strip in what's called a Return to Landing Site or RTLS abort.[18] Instead, the vehicle would go to orbit. "Pretty straightforward," says Young. "A lot less shake-rattle-and-roll than Apollo. You get one and a half Gs off the launchpad and three Gs closer to orbit." One and a half Gs is one and a half times the force of normal gravity, caused by the acceleration of liftoff. Instead of getting lighter and lighter closer to orbit, the force of gravity actually increases. One veteran astronaut described the feeling as having a gorilla sit on your chest.

Americans were flying in space again for the first time since Deke Slayton, Vance Brand, and Tom Stafford blasted off on Apollo-Soyuz in 1975. Bob Crippen was finally circling the earth, something he'd been waiting for since the late 1960s.

For John Young, it was a familiar sight. "Well, you look out the windows once you're in orbit and you can barely keep your head inside the cockpit," he says. "It's so beautiful. You can see the horizon and the earth below."

Young's flight couch was just a couple of feet from Crippen's, but the space veteran couldn't see his rookie crewmate's reaction. He was still in his heavy pressure suit and space helmet. The former astronaut candidate with the Manned Orbiting Laboratory project was on his way to orbit, but not on a Titan 3 rocket with a space station attached. Crippen says he didn't even take his eyes off the dashboard during liftoff since that's where a problem would show up first, in the form of a warning light that flashed or a gauge that looked wrong. "It was obvious that we were losing foam [insulation] off the external fuel tank. Then we pitched over [to settle into orbit] and you could see the earth. It was an exciting view, but I didn't pay much attention until main engine cut off."

Capcom Dan Brandenstein didn't have to see Crippen to gauge his reaction to being in space. He could hear it. "That's when Crip's excitement came through a little," says Brandenstein. "It was his first flight, and when he got a chance to look out the window, the excitement in his voice, from my perspective, went up once they made orbit."

Columbia Goes to the Wrong Orbit

Once the shuttle achieves orbit, the spacecraft adjusts its position with small jet thrusters fueled by two self-igniting chemicals. The thrusters, clustered at the nose and tail of the spacecraft, resemble shower heads. But instead of water coming out of each of the holes, there are alternating spouts of these two corrosive chemicals, which explode on contact. Two larger versions of these thrusters, called the orbital maneuvering system or the "big boomers," slow the shuttle's speed around the earth.[19] That way, instead of circling the globe again, *Columbia* drops down for its planned landing at Edwards Air Force Base. That's when the wing flaps and tail rudder can spell the difference between a successful touchdown and a crash. Pilots consider this part of a shuttle mission especially "righteous."

Astronauts were anxious to trade capsules for wings. But those wings meant aerodynamic lift, and that contributed to problem number two. Young and Crippen went into the wrong orbit. "The wings gave us more lift than we anticipated," says Crippen. "That put us into a different orbit than the one we had canned [programmed]."

The situation became apparent to Mission Control in a hurry. The launch had been carefully rehearsed, and both the astronauts and the

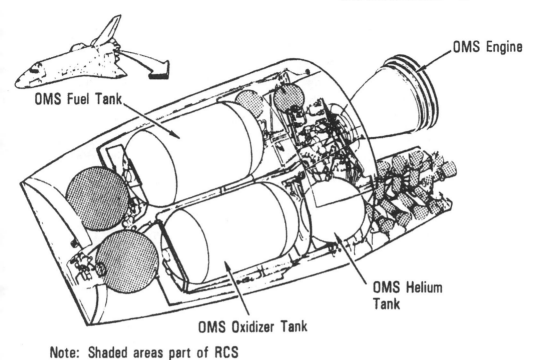

OMS Engine

OMS Fuel Tank

OMS Helium
Tank

OMS Oxidizer Tank

Note: Shaded areas part of RCS

Figure 13. Orbital maneuvering system pod. The orbiter has two of these "big boomers," one on either side of the tail rudder. The large bell-shaped nozzles emit 6,000 pounds of thrust to slow the spacecraft so it can begin the long trip from orbit to the ground at mission's end. Courtesy of NASA.

mission managers had trained on what would happen during the ride to orbit. But if anything went wrong, Dan Brandenstein's cue cards would be tossed aside. "It was only a matter of a few miles," says Brandenstein. The capcom continued his radio messages every thirty seconds or so during *Columbia*'s climb, not so much to provide vital information to the crew as to make sure the lines of communication stayed open on the untested radio system. John Young was equally unruffled by missing the orbit. His job was to test the orbiter and see how it would handle during each phase of the two-day mission. The astronauts blamed the problem on the new solid rocket boosters and the amount of thrust they generated during lift-off.[20] Again, there were no shuttle test flights and no way to measure the performance of the big solids in a wind tunnel. Young and Crippen would have to find out the hard way by "lighting the candle" and making the trip to the wrong orbit. "It's okay," says Young. "During my Gemini and Apollo flights, we never went to the right orbit then either. You just do a [course] correction and you get there sooner or later."

The surprises didn't end there. Young and Crippen looked out the rear windows of the spacecraft's flight deck and saw that several heat protection tiles had torn off from a spot near the tail. Young and Crippen would find out soon enough whether the engineers who held all those meetings were right about tiles coming off the shuttle's belly and holes being burned in the space plane's fuselage.

For Americans, it was the first views of NASA astronauts in space since Tom Stafford shook hands with cosmonaut Alexei Leonov during the Apollo-Soyuz test project in 1975.

It wasn't quite the same, though.

Instead of watching crew members float around inside a gumdrop-shaped Apollo capsule, it looked more like a jetliner. As astronaut John Young sat in his flight seat overlooking his dashboard filled with buttons and gauges, he looked like a pilot on a routine trip from airport to airport on the earth, except that objects next to him were floating gently in the absence of gravity. Bob Crippen did somersaults in the lower level, or middeck, of the shuttle. Behind him were row upon row of white lockers, giving the orbiter's interior a sterile appearance sometimes compared to that of a hospital. The view out the double-paned windows on the upper level, or cockpit deck, was different as well. Once *Columbia* was in orbit, Young and Crippen opened the huge clamshell doors enclosing the cargo bay, exposing the area behind the crew cabin for payloads as big as the large spy satellites favored by the Pentagon. This is how many TV viewers on the ground were introduced to the delta-winged vehicle with the cargo capacity NASA needed to win the participation of the air force.

After orbiting the earth for two days, the crew put their pressure suits on again for the glide to Edwards Air Force Base in California. First there was the reentry through the earth's atmosphere, which began with firing the shuttle's biggest jet thrusters. Slowly gravity returned in the crew cabin. Shuttle pilots later would gauge the progress of this stage of reentry with what they called the pencil test. They set a pencil in front of them and watch it hang there motionless. As the spacecraft comes under the influence of gravity, the pencil slowly begins to drift down toward the floor.

Another clue that reentry was occurring can be seen outside the windows as the orbiter's black belly tiles scrape against the atmosphere, generating 3,000 degrees of heat and a fire show for the crew. "We started reentry on the night side of Earth," says Crippen. "You get this pink glow outside the windows. I compare it to flying in a neon tube." It would be the white-hot gases of reentry that blew past a broken heat shield on *Columbia* like a blowtorch in February of 2003. The cracked shield on the leading edge of the left wing and the extra drag on the shuttle forced the vehicle's autopilot to struggle to stay on course. The damage grew, and NASA's first space shuttle disintegrated at 10,000 miles per hour just sixteen minutes from its intended landing at the Kennedy Space Center. That tragedy was years after *Columbia*'s first landing in 1981, but the tiles were a factor back then as well. Engineers expected the mosaic of ceramic tiles to generate drag on the orbiter. It didn't, and the spacecraft had more lift than the astronauts anticipated as the spacecraft glided down to the dry lake bed at Edwards.

The sight after *Columbia*'s touchdown was also different from the way Young came home from his 1971 Apollo moon mission. Back then, the command module capsule nicknamed Casper plopped down into the Pacific Ocean by parachute. The Apollo astronauts were then intercepted by the aircraft carrier USS *Ticonderoga* and brought back to the United States. When *Columbia* rolled to a stop in 1981 on runway 23 at Edwards Air Force Base, it was a met by a more modest convoy of trucks. The ground crew wheeled up a set of stairs as if they were greeting a pampered passenger jet, compared to a space capsule plucked from the sea after reentry. After an hour, Young and Crippen descended and started shaking hands. Onlookers say Young looked giddy as he checked out the vehicle. The engineers were wrong, the belly tiles were still there, and he and Crippen were still alive.[21]

Critics say the space shuttle was the result of too many cooks spoiling the broth, and a lot of people were wearing a chef's hat as NASA put the program together. Still, whether it had a mission or not, the maiden launch of *Columbia* on STS-1 was deemed a success, and America was back in space.

4

1982–1985

The Curse of Being Routine

NASA was basking in the success of STS-1. Landing the spacecraft at Edwards Air Force Base in California's vast Mojave Desert meant having to ship *Columbia* back to its Florida launch site aboard a modified 747 jumbo jet so it could be processed for its next flight. This gave NASA an extra chance to advertise its new achievement by making refueling stops at several air force bases between California and Florida where news crews could get a look at the winged vehicle bolted to its carrier jet.

Despite the favorable headlines, debate still raged among NASA's detractors on what the shuttle was supposed to accomplish, without a clear objective to work toward. Critics like former air force secretary Robert Seamans persisted in asking why the agency should risk astronauts on missions to launch commercial satellites when unmanned rockets could do the job. NASA justified the shuttle by citing the things humans could do in orbit, like react to emergencies and make repairs. The shuttle's ability to carry heavy payloads back to Earth was also touted, but even NASA's future administrator Michael Griffin admitted before taking office that this "down cargo" capability was seldom used.[1]

The astronauts who lined up to follow Young and Crippen into orbit on the shuttle would not be placed in the same limelight by the media. This situation was common in the space program. The press referred to it as the Apollo 12 Syndrome.

The world was on the edge of its collective seat as the astronauts of *Apollo 11* headed to the Sea of Tranquility for the world's first manned moon landing in 1969. But ask your friends, do they remember the Ocean of Storms? That was the landing spot for *Apollo 12*. A similar feeling of lonely anonymity dogged the second one-man Mercury flight in 1961, the second two-man Gemini mission in 1965, and the second flight of the space shuttle. That was STS-2, set for blastoff in November of 1981.

Launch engineer Sam Beddingfield was happy that *Columbia*'s first launch turned out okay. But he was also adding up everything that went wrong, and there were some disturbing problems that needed to be resolved before the vehicle could go again. One was the explosion that occurred at the launchpad. When supercold fuel is pumped into *Columbia*'s white external fuel tank, some of it warms up and forms what looks like steam coming out of a vent at the top of the tank.[2] That extra hydrogen collected at the bottom of the launchpad, and when the three liquid-fueled engines of STS-1 ignited, the hydrogen exploded near the shuttle. "It bent the wing flaps on the shuttle," says Beddingfield. "They were really out of position. If they bent much more, they might have broken off." That could have meant a catastrophe before the mission got under way.

The lives of the shuttle astronauts would depend on those flaps, at the end of their missions on the new vehicle. Following blastoff, the orbiter is basically an unpowered glider. Beddingfield's job was to help make sure the shuttle got off the pad in one piece the next time. The cause of that launch-day explosion needed to be resolved before the next liftoff. Astronauts were training for their chance at a spaceflight following the end of Project Apollo, but *Columbia* might not survive many more launches if nothing was done.

The problem for Beddingfield was how to get rid of the hydrogen in the seconds before liftoff when no pad workers could be anywhere near the shuttle.

The answer was found at a local dime store. "It's the same stuff you get when you buy sparklers," says Beddingfield. The point was to burn away the hydrogen. The image of sparks sputtering around the engine nozzles is now part of the routine of launching a shuttle. In 1981 it was just an idea on which Beddingfield had to follow through in a hurry. About that time, local stores still had their stocks of Fourth of July sparklers. That set local kids, and at least one NASA engineer, thinking about the possibilities.[3]

The result—the igniters or flares that became a standard part of the spectacle of launching a shuttle—officially goes by the name *space shuttle main engine hydrogen burnoff system*. The main engines start at six seconds before zero. The flares on either side of the engine nozzles go off like roman candles at ten seconds prior to zero.[4] STS-2 would be the first launch to utilize the igniters.

Old Skills Become New Again on STS-2

Columbia's second commander, Joseph Engle, originally trained as an air force test pilot in the 1950s and was picked as one of the dozen pilots to fly the X-15 rocket plane, built by the contractor North American Aviation. NASA teamed up with the navy and the air force for test flights starting in 1959. The X-15 remains the fastest aircraft ever built. There were three of the black, stubby-winged vehicles, and Engle would be one of the men taking turns flying them. Another pilot chosen for the X-15 program was a native of Wapakoneta, Ohio, named Neil Armstrong.[5]

The flights were designed to test how men and aircraft would function at the very edge of space. A B-52 bomber would carry the rocket plane to an altitude of 45,000 feet and cut it loose, and then the X-15 and its pilot would "haul the mail."[6]

The X-15s would make a steep climb to as high as 300,000 feet at speeds as fast as six times the speed of sound. At that altitude, the vehicle's rocket engines would have to be augmented by reaction control system thrusters to give the pilot control. The chemically fueled RCS jets would later find their way onto NASA's space shuttles. Getting pilots ready for these demanding sorties prompted the use of computerized cockpit simulators, which would also come into play during Apollo and the shuttle program. The X-15 pilots were studied as much as the aircraft. Stress levels due

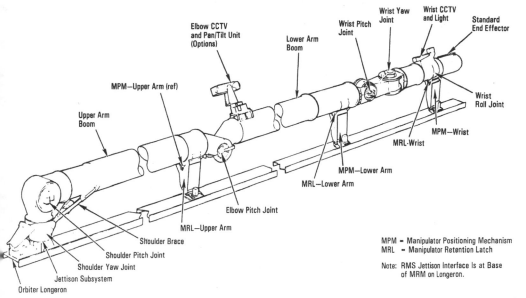

RMS—*Stowed Position and General Arrangement*

Figure 14. Robotic arm. The fifty-foot-long arm, built by Canada, is attached inside the shuttle's cargo bay. Astronauts use it to grapple cargoes or to grab objects in orbit after the shuttle effects a rendezvous. A later version for the International Space Station is free floating and has grappling mechanisms on either end to allow the device to "inchworm" its way around the outpost or ride a railcar along a track on the outside of the complex. Courtesy of NASA.

to the speed and altitude, not to mention the plummet toward landing, would push heart rates and blood pressure to dangerously high levels. The program didn't have a perfect safety record, with two crash landings and one incident where an X-15 was blown in half due to a mechanical failure.

The applicability of the X-15 experience to the later space shuttle program was easy to see. Both craft had to operate at extreme speeds and in the hostile environment of space. The landings were also similar, since the stubby-winged spacecraft made a steep final approach as a glider with one shot at a safe "dead stick" touchdown.[7]

Even though *Columbia* was a more contemporary vehicle, the X-15 was more maneuverable and was versatile enough to cover a wide range of test objectives. The shuttle was a comparatively lumbering type of aircraft designed for a specific flight pattern. The X-15 also tested the kind of steep dive-bomber approach to landing that would be standard on the shuttle. The X-15 program helped give NASA confidence that *Columbia* could do it as well.

The job of Engle and copilot Dick Truly on STS-2 was to pick up where Young and Crippen left off. The list included gathering more data on how *Columbia* handled during liftoff, while it orbited the earth, and during the critical landing. STS-2 also included the first in-orbit tests of the fifty-foot-long robotic arm. The remote manipulator system, built by the Canadian Space Agency, would go on to grapple satellites and to deploy others like the Hubble Space Telescope. The absence of a robotic arm on *Columbia* during its final, fatal flight in 2003 became a major issue for NASA, since the astronauts were unable to inspect their vehicle for damage.

The blastoff occurred on November 12, 1981. The flight was supposed to last five days, but the failure of an onboard fuel cell would force mission managers to cut it to two. The fuel cells on the shuttle generate electricity and, as a by-product, drinking water.

After two more test flights, the shuttle program would be declared ready to move on to deploying satellites and carrying extra crew members. The inclusion of more than a commander and a pilot made the use of jet-fighter ejector seats impractical. Unlike Mercury, Gemini, and Apollo capsules, the shuttle didn't have a crew escape system for crews larger than two. Future astronauts would have to take their chances in case of a problem.

Perhaps the most disquieting result of STS-2 was the damage that was discovered on the primary O-ring gasket on *Columbia*'s right-hand solid rocket booster. Each of the two rockets is made up of cylindrical segments filled with solid fuel that burns nonstop once the booster is ignited. The segments are stacked one on top of another like a column of soft-drink cans. A pair of O-rings, a primary and a backup, are packed inside each joint between the segments, and then asbestos putty is applied to the inside of the seal. On STS-2, gaps in the putty were blamed for allowing booster exhaust to burn through to the primary O-ring and begin to erode it. A complete burn-through didn't occur, but the O-ring damage was the worst in the history of the space shuttle program, with one exception. The only vehicle to suffer a worse problem was *Challenger* on mission 51L in

1986. That mission, of course, ended with the disaster that killed all seven astronauts on board, including teacher-in-space Christa McAuliffe.[8]

NASA's First Nighttime Space Shuttle Launch and Landing

After STS-2, NASA would go on to make two more test flights of the space shuttle before declaring the program "open for business" and ready to accommodate paying customers. *Columbia*'s next flight, called STS-5, carried up two communications satellites for a Virginia corporation and for Telsat Canada. In addition to the pilot and the commander, who would control the shuttle during the critical landing, there were two more astronauts, called mission specialists. This new breed of shuttle crew member was trained to operate systems on the spacecraft like the robotic arm, which had undergone checkout runs during the earlier test flights of *Columbia*. These astronauts would also be ready to put on bulky spacesuits and float outside on spacewalks, either to conduct experiments or to make repairs or retrieve a satellite cargo. Even the standard crew photograph taken before launch signaled a change of attitude in the shuttle program. Instead of the rugged-looking tan pressure suits worn during the four test flights, the STS-5 crew were clad in flight suits of powder blue cloth with pleated shoulders, giving a more futuristic look. It wasn't just a fashion statement. NASA stopped using ejector seats with this flight, so the protection of pressure suits was considered unnecessary. The *Columbia* crew would wear clamshell helmets to provide a steady flow of air, and that's all. The trend would continue until the *Challenger* disaster in 1986. Then the heavy pressure suits would come rapidly back into style.

Despite selling the shuttle as a launch service with the added bonus of astronauts, neither spacewalker Joe Allen nor Bill Lenoir could have done much for either satellite in *Columbia*'s cargo bay. A malfunction in one of the spacesuits prevented the first spacewalk outside the shuttle. Later flights would carry up a NASA communications satellite and one built by Germany before trying something more ambitious. Up to now, all of the launches and landings of *Columbia* and the newest shuttle, *Challenger*, were during the daytime so the astronauts and Mission Control could see where the spacecraft was going. For the eighth mission, the shuttle would launch and land at night.

Dan Brandenstein, capcom for John Young and Bob Crippen during STS-1, would be assigned to the new flight crew. Now another member of the astronaut corps would be talking over the radio to Brandenstein's crew in space as he had done during the inaugural shuttle flight. One of the cargoes on STS-8, set for launch in late August of 1983, was a boxy satellite covered in gold protective foil called INSAT 1-B, for the nation of India. "That's why we went for a nighttime launch," says Brandenstein. "We needed to ride to a spot to launch [INSAT] so, when all was said and done, it wound up over India."

Launching at night was one hurdle, but the perilous landing in the dark would take extra training for the *Challenger* crew. The added preparation was needed so the astronauts could align their spacecraft for touchdown during the plunge from orbit to the dry lake bed at Edwards Air Force Base in California's Mojave Desert, where both the sky and the fast-approaching runway were pitch black. "There are navy people who put optical landing systems on aircraft carriers, and they came out and helped us," recalls Brandenstein. "We looked at a variety of ways to light the runway, and we had to develop something that worked during the glide slope."

There was no question but that the first nighttime landing would occur at Edwards. Touchdowns in Florida at the Kennedy Space Center's 15,000-foot concrete landing strip wouldn't begin until five months after STS-8, and during daylight hours. The strip at KSC is tightly surrounded by trees, swamp, and the Intracoastal Waterway. NASA wouldn't even consider landing in the dark at Kennedy until a mission by shuttle *Discovery* in 1991. That crew was diverted back to Edwards with "bad weather" given as the reason. Another opportunity didn't come for a nighttime KSC landing until 1993.

While STS-8 would make the more conservative Edwards landing, that didn't mean NASA wasn't ready to push the envelope during launch. At least, that was Dan Brandenstein's impression. He was strapped into the pilot's seat on *Challenger* alongside commander Dick Truly, who had flown *Columbia*'s second mission. The rest of the crew was made up of astronauts Bill Thornton, Dale Gardner, and Guion Bluford, the first African American to go into space.

The weather looked bad on launch night, with thunderstorms in the area. Brandenstein thought they would scrub the liftoff. But then word came from astronaut Bob Crippen, who was flying a NASA training jet to get a firsthand look at the rain situation. "It was raining like crazy," Brandenstein remembers. "With the launch criteria we have now, we'd never have launched. But Crip made the call, and we went just seventeen minutes late. It was a quiet ride, like you were flying inside a fireball. All you could see was the glow from engines around the cockpit."[9]

Nighttime launches are visually spectacular and, up to that point, were somewhat rare for NASA. Private pilots flying small aircraft outside the restricted zone around the launchpads describe blastoff as lighting up most of the Kennedy Space Center.

The liftoff of STS-8 confirmed an earlier observation by Dan Brandenstein that you could enjoy a launch better from Mission Control as capcom than strapped in the cockpit. The rookie astronaut's attention was focused on the orbiter's systems to make sure they were working. The first inkling that he was really in space came after blastoff, after the boosters had separated, and after the main engines had cut off. "Coming up over Africa," Brandenstein says, "we hit the sunrise and then we were in orbit. So that was my first visual from space. Of all my times in orbit, that sticks in my mind as the most memorable."

The STS-8 astronauts also commemorated their flight with an extra crew patch, beyond the tradition of creating a patch to symbolize a particular mission and wearing it in space that dates back to the *Gemini 5* flight of Gordon Cooper and Pete Conrad. The crew patch featured a Conestoga wagon and the slogan "Eight days or bust," referring to the marathon nature of their flight. Older and rarer versions of the patch, sought by space memorabilia collectors, lack the slogan.[10]

The official patch pictured the shuttle during liftoff and the name *Challenger* across the center. The extra crew patch from STS-8 was meant as a joke. It was mostly black with just the eyes of the flight crew visible, peering out the windows of the shuttle. Brandenstein's experience flying A-6 Intruder aircraft for the U.S. Navy was the inspiration for the black patch. Crews on this type of aircraft often flew night sorties and had black patches made to mark the occasion. The only point about the STS-8 patch that space observers debate is, which eyes are whose? Four of the sets of eyes look scared, and the fifth is depicted wearing eyeglasses. One story is that the man in glasses is commander Richard Truly; the other is that it's astronaut William Thornton getting ready to draw blood samples from the crew, and the look of fear is due to the needle. Brandenstein says the Truly story is the right one. "That was the deal," he says with a smile in his voice. "Dick Truly was the stern-eyed veteran making the nighttime landing, and we were all the wide-eyed rookies riding along."

And then there were the souvenir envelopes.

NASA was marking its twenty-fifth anniversary, and Truly suggested carrying along 160,000 envelopes bearing the STS-8 mission patch. The $9.35 postage stamp, printed with the portrait of an American bald eagle, would be stamped with three official post office cancellations to certify that the envelope had indeed flown in space. One was from the Kennedy Space Center for August 14, the original launch date. The second was another KSC cancellation dated August 30, the day *Challenger* actually blasted off. The final stamp was from Edwards Air Force Base in California, dated September 5, the day *Challenger* rolled to a stop at mission's end. NASA and the U.S. Postal Service cooperated on the stunt—and, in the process, resurrected a ghost from the agency's Apollo days.

Astronauts routinely carry souvenirs in orbit for themselves, for their friends and families, and for supporters of the space program. The list ranges from footballs and soccer balls to embroidered patches to the cremated remains of *Star Trek* creator Gene Roddenberry.[11] Astronaut Alan Bean left his silver rookie pin on the moon. The agency drew the line when *Apollo 15* astronauts David Scott, Alfred Worden, and Jim Irwin carried four hundred onionskin envelopes to the surface of the moon. On their return, the agency declared the action improper and impounded the envelopes. Years later, when NASA endorsed the inclusion of the commemorative envelopes on STS-8, one *Apollo 15* crew member asked for the moon envelopes back, and NASA conceded the point.[12] Collectors have been known to pay prices in the five figures for an *Apollo 15* envelope, while one of the thousands of STS-8 commemoratives can now be found on eBay for just a few dollars.

Astronaut John Young's Next Mission, and His Last

With the successful conclusion of STS-8, NASA moved on to the first of a number of missions featuring the European-built Spacelab. The international venture frequently included international crews with payload specialists from Germany, Japan, Canada, and the Netherlands, among others. The school-bus-sized Spacelab was nestled into the shuttle's payload

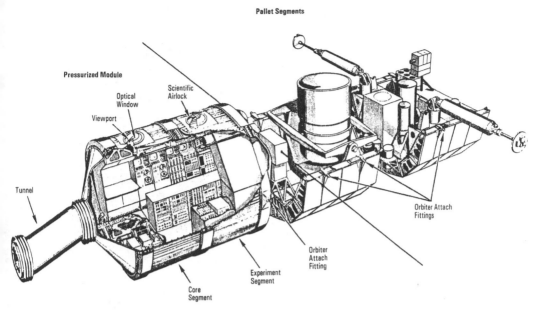

European Space Agency's Spacelab

Figure 15. Cutaway view of Spacelab. The European-built facility first flew aboard space shuttle *Columbia* in 1983. A pressurized module the size of a railroad car afforded scientists a shirtsleeve environment for experiments in materials science, the biological sciences, astronomy, and other disciplines. Its final flight was Neurolab in 1998. Courtesy of NASA.

bay, with a tunnel connecting the module to the crew cabin. Astronauts would float from their cramped living compartment to the much roomier Spacelab, where racks of equipment held glove boxes, furnaces, freezers, and even cages for lab animals like monkeys and rats.

NASA's request that the European Space Agency build Spacelab also meant the shuttle could utilize the full volume of the vast cargo bay ordered by the air force. The very first Spacelab flight would carry a record six crew members. The lab and the astronauts would fly on *Columbia*, and astronaut John Young would be commander on his final shuttle flight. The launch was supposed to be in September of 1983, but a problem with one of the solid rocket boosters forced a rollback.

The spacecraft are maintained at the Kennedy Space Center in hangars called Orbiter Processing Facilities near the box-shaped Vehicle Assembly

Figure 16. Crew portrait of STS-9, the first crew to work inside the European-built Spacelab module. Front, left to right: astronauts Owen K. Garriott, Brewster Shaw, John Young, and Robert Parker. Back, left to right: astronaut Byron Lichtenberg and ESA astronaut Ulf Merbold. Courtesy of NASA.

Building. Once a shuttle is declared ready for flight, technicians wheel it over to the VAB, where it's bolted to its external tank and twin boosters. The whole vehicle, called the launch stack, is loaded onto one of two giant tractors called crawlers for the slow roll to the pad. The crawlers are leftovers from Apollo, as are the launchpads themselves. All that equipment was built to accommodate the massive Saturn V rockets that carried men to the moon. They were later passed as hand-me-downs to the shuttle program.[13]

STS-9 was the first time a shuttle poised on the pad was rolled back to fix a problem. Blastoff occurred in November of 1983, and John Young found himself managing a ship full of scientists as well as the shuttle itself. "It worked out very well," he remembers. "We worked twelve-hour shifts in teams, and we'd bump into each other during shift change. We'd have breakfast and supper together and then go our separate ways."

The crew included rookie pilot Brewster Shaw, Skylab veteran Owen Garriott, and German astronaut Ulf Merbold representing the European

Space Agency. Perhaps the one most at home in the roomy Spacelab compartment was Garriott, who had spent almost sixty days orbiting the earth in the much larger Skylab space station in 1973.

The group broke up into teams of three, one led by Young and the other by Shaw. "Brewster and I would do the shuttle stuff," says Young, "like changing attitude and orbit, and the other two guys would work in the back. Spacelab had, like, seventy experiments, and we got 'em all done." The flight had its share of problems, though: two of *Columbia*'s five main computers broke down, a navigation box was lost, and two of the hydraulic units caught fire during landing.[14]

For a time, NASA stopped calling missions STS-1, STS-2, STS-3 and utilized a more arcane system including letters. Here's the code breaker. STS means Space Transportation System. In astronaut Joe Engle's next mission, called STS-51I, the 5 meant 1985, the fiscal year in which the mission was meant to fly; the 1 meant Kennedy Space Center was the launch site; and the letter I meant that Engle's flight was the ninth mission set for that year, with I being the ninth letter in the alphabet.

Dan Brandenstein Loses a Crewmate
to a Changing Schedule and *Challenger*

The year 1985 was also marked by a lot of flight shuffling and confusion for the astronauts. Launch managers cancelled some missions and reassigned crew members and remanifested cargoes. That happened to Dan Brandenstein during his first flight as commander, on STS-51G, one of NASA's most international flights to date. The astronauts deployed three satellites, for AT&T, Mexico, and Saudi Arabia. A Saudi prince was assigned as a payload specialist to observe for the royal family. NASA breaks down shuttle crews in categories. Commanders are in charge of the spacecraft and mission, pilots assist the commanders and help fly the shuttle, mission specialists are trained astronauts with specific skills needed to operate systems on the spacecraft, and payload specialists are scientists or observers who operate a specific set of experiments. This shuttle flight would include astronauts John Fabian, Shannon Lucid, John Creighton, and Steve Nagel, along with payload specialists Patrick Baudry of France and Sultan Salman al-Saud.

Figure 17. Crew portrait of shuttle mission STS-51G. Front, left to right: astronauts Dan Brandenstein and John Creighton. Back, left to right: astronauts Shannon Lucid, Steve Nagel, and John Fabian, Saudi prince Sultan Salman al-Saud, and French payload specialist Patrick Baudry. Courtesy of NASA.

For Dan Brandenstein, the inclusion of a Saudi prince and a French scientist meant extra training for the crew. "We had a group of people from Aramco in Houston come in," says Brandenstein. "They worked a lot of the oil fields in Saudi Arabia, and they spent a day with us and gave us some background in the [Saudi] culture." NASA later found it was unnecessary, since the prince was educated in the United States and proved to be as Americanized as any member of the *Discovery* crew.

The mission was uneventful, but one aspect continues to haunt Brandenstein. The original version of this STS-51G mission was called STS-51D and had a different list of objectives. That crew would have deployed NASA's second Tracking and Data Relay Satellite to beam pictures and data from shuttles or other spacecraft to Earth. Another planned task was retrieving a boxcar-sized experiment package called the Long Duration Exposure Facility. Astronauts had left LDEF in orbit the year before, and

Brandenstein's crew was to bring it back from space. NASA cancelled that mission and moved two payload specialists to other missions. One was a scientist from Hughes Aerospace named Greg Jarvis, with whom Brandenstein worked closely.

"We had our L-minus-30 press conference," says Brandenstein. "Then, that afternoon NASA called us in and cancelled the mission." The L-minus-30 event with reporters occurs, as the name implies, thirty days before liftoff. Brandenstein and his copilot John Creighton were assigned to the new STS-51G mission along with astronauts Nagel, Lucid, and John Fabian.[15]

NASA had other plans for Greg Jarvis. He was told to join the crew of shuttle *Challenger* on mission STS-51L, which would include teacher-in-space Christa McAuliffe. Launch was set for January of 1986.

5

1986

"Don't Call Unless It Blows Up"

The directive "Don't call unless it blows up" became prophetic on the morning of January 28, 1986, the morning *Challenger* lifted off for the last time. It was a terse response from an editor to a radio reporter stationed at NASA's Kennedy Space Center who called to pitch a news item about the impending blastoff. The twenty-fifth launch of the space shuttle wasn't seen as much of a story that day. The shuttles had been flying for almost five years, and the media considered the launches less and less newsworthy as each successful trip made the program seem routine. Even the inclusion of the first "teacher in space," complete with Christa McAuliffe's broad smile and all-American looks, failed to generate a lot of interest. It would take something spectacular to put this blastoff into a newscast, and the shuttle's explosion would do so at 11:39 that morning.

Up to now, instead of pursuing a specific mission, NASA had been using the shuttle to launch satellites and to allow the astronauts to practice what might be done on future, more ambitious ventures. For example, spacewalkers floated outside the shuttle in heavy spacesuits to go through the motions of refueling spacecraft in orbit. Two crewmen on a later flight tinkered with long girders that might be assembled by hand to form the struts of a yet-to-be-decided-upon space station.

Shuttle flights had also become the ultimate congressional junket. U.S. senator Jake Garn of Utah flew aboard shuttle *Discovery*, which was notable for blowing a tire upon landing back in Florida. Congressman Bill Nelson of Florida blasted off aboard *Columbia* on the last mission to fly before NASA, and America, learned how risky the shuttle program actually was.

Still, despite the repetitive nature of the launches, NASA had its fans during this time.

As the agency prepared to launch *Challenger* on shuttle mission STS-51L in early 1986, a photo of the astronauts, in their blue flight suits, was tacked to the bedroom wall of nine-year-old Jenny Eschen of Floral City, Florida. The diverse crew had someone with whom almost any school student could identify. Little girls like Jenny could follow the experiments and lessons conducted by astronaut Judy Resnik and teacher Christa McAuliffe. Mission specialist Ron McNair was an African American who had taken his saxophone into space on his first flight in 1984. Ellison Onizuka was from Hawaii and had flown aboard shuttle *Discovery* on a secret Defense Department mission in 1985. Rookies Mike Smith and Greg Jarvis were making their first launch, and veteran Dick Scobee would lead the crew.

The morning of January 28, young Jenny and a busload of her friends and classmates were bundled up for the cold ride across the state to the Kennedy Space Center to watch the liftoff. It was 27 degrees Fahrenheit out at the launchpad. The effect of the cold weather on the O-ring gaskets on one of *Challenger*'s booster rockets would later be blamed, in part, for the shuttle disaster.

When she was interviewed almost two decades after the accident, Jenny Eschen—now known by her married name, Jenny Carter—said she had managed not to think about *Challenger* for a few years. Then there was a flood of memories, which welled up for her and her family during their Christmas celebration. "It's like, where were you when Kennedy was shot?" Jenny says. "So, where were you when *Challenger* exploded? I mean, everyone remembers." Jenny was years from the start of her ca-

Figure 18. Crew portrait of *Challenger* astronauts on mission STS-51L. Front, left to right: astronauts Mike Smith, Dick Scobee, and Ron McNair. Back, left to right: astronaut Ellison Onizuka, teacher-in-space Christa McAuliffe, payload specialist Greg Jarvis, and astronaut Judith Resnik. Courtesy of NASA.

reer in banking in Atlanta when she made the three-hour trip aboard an unheated school bus over the rolling hills of Citrus County and out to the flat Brevard County terrain that is home to the Kennedy Space Center. "Back then, I was a science geek," she says, "so *Challenger* was a huge deal for me."[1] Along with lessons from orbit by New Hampshire teacher Christa McAuliffe, the *Challenger* astronauts planned to launch a NASA communications satellite to help transmit data from space to the earth. The network would eventually exchange grainy black and white images of astronauts for the dramatic color shots from the Hubble Space Telescope and the International Space Station that NASA observers would later find commonplace. Also, 1986 was the year that Halley's comet would make its first visit close to Earth since 1910. *Challenger*'s cargo included a robot satellite called Spartan-Halley, which would study the icy comet during a brief mission on its own before the astronauts would retrieve it and lock it in the cargo bay with its trove of data to be downloaded after landing.[2]

On launch day, the bus carrying nine-year-old Jenny Eschen and her classmates rattled up to the main gate of the Kennedy Space Center just minutes before liftoff. The gate was on a causeway linking the Florida mainland to the chain of barrier islands on the state's Atlantic coast that includes the land where KSC is located. Across the Banana River, the box-shaped Vehicle Assembly Building was visible to Jenny and her friends. Just to the right was *Challenger* on launchpad 39B. Veteran NASA engineers say two pads were created in case of a disaster at the moment the shuttle's engines ignited. They admit, though, that the way *Challenger* exploded caught them by surprise.

Jenny and her classmates stood poised at the main gate of the Kennedy Space Center. She could hear the countdown through loudspeakers set up by NASA for visitors not lucky enough to get a seat at the official reviewing stands. Cars, buses, and recreational vehicles lined the Florida coast at prized spots like Jetty Park in the town of Titusville, which sits across the Intracoastal from KSC and affords a clear view of the launchpads and the shuttles. Jenny Carter remembers the sights and sounds as the seconds ticked down. Tight in her hands was the camera her parents gave her for Christmas weeks before. "It was just a little kiddy camera," she said. "Everybody said just keep clicking, just keep clicking." She still has those photos in her scrapbook. The shuttle is clearly visible, atop a column of flame as the space plane leaped off the pad at 200 miles per hour.[3]

Inside the space shuttle, the crew members were in their blue cloth coveralls and space helmets, strapped in their flight seats for the launch. On the flight deck, commander Scobee and pilot Smith sat behind the cockpit control sticks and instrument panels. Behind them was astronaut Resnik, who would act as flight engineer in support of the pilots during liftoff. To her right, up against the bulkhead, was Onizuka. They had the best view during launch with the large windshields directly above them. The rest of the crew sat on the orbiter's lower deck, or middeck, where the view was more limited. No big windows, but this was where the crew would eat and sleep while in orbit. The front wall was also lined with lockers filled with equipment for the mission. That's where veteran astronaut Ron McNair was seated, next to the round main hatch of the spacecraft. Next to him were two first-timers, McAuliffe and Jarvis.[4]

While the astronauts waited through the final minutes of the countdown, the kids from Citrus County were having a great time waiting for the liftoff of *Challenger*. It was a great time for a while, anyway. "I saw the two boosters go off in two directions," said ten-year-old Robby Samson after the accident. "You knew there was a problem, especially when there was that big cloud of smoke."

"It was really bright, and then you could see them [the boosters] go off in two directions," said ten-year-old Jennifer Golden. "You could tell there was a problem."

Nine-year-old Jenny Eschen had this observation: "I was expecting a safe flight, because I thought it was exciting to have a teacher in space." That notion was over. A huge mushroom cloud hung in the sky over the Kennedy Space Center as the kids from Floral City Elementary were herded by security guards. The teachers supervising the field trip to the Kennedy Space Center had the job of getting the kids out of there. "People started appearing with barricades," says Teresa Manning. "They said, 'You have to leave the Cape now!'" Manning, who taught the gifted program at Floral City Elementary School, had accompanied the teacher of Jenny Carter as a chaperone. Following a three-hour trip in a chilly school bus to get to the Kennedy Space Center, they and the kids faced turning around and making the three-hour trip back without fully understanding what they had seen. "The children on the bus were very quiet the whole time," remembers Manning. "We had a radio on the bus, and we tried to get news reports so we could talk to the kids."[5]

Jenny Eschen's parents were on edge.

"It was like 9-11," Jenny says. "My dad was teaching at school, but my mom was at home in front of the television and they just played the explosion on the news over and over and over again." This was 1986, a time before cellular phones. The bus carrying Manning along with Jenny and her schoolmates made a lunch stop at a McDonald's restaurant. There was a pay phone available, so Manning could tell the school what happened and that the kids were okay. Not everybody got the message. Jenny Eschen got home just after six. Her parents were there and the video images of *Challenger* exploding were still playing on TV. "My parents didn't know what was going on," she says. "They didn't know if I got hit by a piece of the shuttle or anything. So, when I came in, they just went bananas. That was really scary."

The national reaction to the *Challenger* disaster was scary for NASA as well. Reporters who originally ignored the teacher-in-space mission now flooded to the Kennedy Space Center. The KSC Press Site auditorium was set up to accommodate the arrival of Vice President George Bush and U.S. senators Jake Garn and John Glenn.

The mushroom cloud from the explosion hung over the Cape for hours. Photography of the accident showed the boosters flying away from the blast, as well as one of the shuttle's wings and the crew cabin containing the astronauts. NASA estimates that lighter pieces, like the heat tiles, kept falling for up to an hour after the shuttle was gone.[6]

NASA Prepares for the Post-Challenger Assault by the Press

The KSC Press Site, occupied by only a handful of reporters before *Challenger*'s launch, was now standing room only. News crews crowded into the facility, grabbing interview sound from anyone with a NASA badge. Inside, a semicircular counter separated the few NASA media specialists from the throng of reporters digging for every fact that could be found.

The fate of the seven astronauts, including teacher-in-space Christa McAuliffe, is well known now. The day of the disaster, NASA officials held out a glimmer of hope that the crew might have survived. "We're still hoping that the orbiter could be somewhat intact and could float and that the crew could be fine," said NASA media representative George Diller during an initial interview. KSC quickly became the epicenter of the biggest space-related news story since the first shuttle blastoff. Just days before the accident, the seats at the podium inside the Press Site auditorium were occupied by obscure air force weather forecasters and NASA scientists no one would ever remember. Now Vice President Bush and Senators Glenn and Garn were there consoling the nation following the deaths of seven astronauts. Bush's comments were aimed at children like Jenny Eschen, Robbie Samson, and Jennifer Golden who were coping with the tragedy. "You must understand that spirit, bravery, and commitment are what make not only the space program but all of life worthwhile," said the vice president. "We must never, as people in our daily lives or the nation, stop exploring and stop discovering."[7]

While the press were gathering the bits of pieces of information to write their stories, NASA was also gathering the billions of bits of telemetry data radioed to ground stations from the shuttle. That information contained a stream of pressures, temperatures, and other readings from the spacecraft's onboard systems. A clue might be hidden there. Speculation on what happened was also just beginning. Some of it was off the mark, but some was eerily accurate.

"We just don't know what happened," said Diller. "It was the first launch from pad 39B—could that have been a factor? There was subfreezing weather—did that have any bearing on it?"

After the *Challenger* disaster, there were answers to be sought as to the cause of the accident. Also, NASA would face serious questions on how it ran the shuttle program and how things could go so terribly wrong.

6

The Investigation

Challenger shattered a key misconception about the space shuttle. While its critics and supporters debated its usefulness and its lack of a central mission, the orbiter wasn't considered particularly hazardous to fly. Now, it was.

That issue served to reframe the debate over the shuttle. With no exact mission to fulfill, the question before the loss of *Challenger* was, why spend tax dollars this way? Following the accident, NASA prepared to answer why it was necessary to risk the lives of seven astronauts on a shuttle flight featuring little more than the launching of two satellites, some science experiments, and several lessons taught by a teacher from orbit. Charges of miscommunication and mismanagement would also result from the official inquiry into the explosion.

The White House assembled a panel to investigate the *Challenger* disaster, under the direction of former secretary of state William Rogers. The onetime advisor to President Nixon during the triumph of the *Apollo 11* moon landing in 1969 would lead the probe into NASA's darkest hour in 1986. Luminaries who would preside over public hearings into the fatal space mission included the first man to walk on the moon, Neil Armstrong; the first American woman in space, Sally Ride; and Nobel Prize–winning physicist Dr. Richard Feynman. The list of candidates for the panel left some observers concerned, since many of the people being recommended had direct or indirect ties to NASA. This raised questions of objectivity as the group prepared to put the shuttle accident, and the U.S. space program, under what was supposed to be an unrelenting microscope.[1] Still, the list went to President Reagan and was quickly approved.

The group was called the Presidential Commission on the Space Shuttle Challenger Accident, but to the public it was known as the Rogers Commission. The White House mandate that created the group on February 3, 1986, set specific goals. The panel was supposed to investigate the cause of the *Challenger* accident through the examination of clues and the taking of testimony. In addition to physical evidence from the disaster, which included a good deal of wreckage, there was telemetry—billions of bits of computer data beamed to Earth by the shuttle during its short flight and even in the moments after the explosion. All of that information would paint a picture of how *Challenger* was performing from its takeoff to its disintegration.[2] The commission would also hear sworn testimony from more than 160 witnesses from NASA, industry, and elsewhere during public hearings. Finally, the panel would publish recommendations on how to fix the cause of the disaster. The commission's final report would excoriate the U.S. space agency and introduce several new words into the American vocabulary—things like "field joint" and "joint rotation" and "O-ring."

But the panel couldn't begin to right the wrongs that doomed *Challenger* without sifting the mountains of data from the scene of the accident, the Kennedy Space Center and the waters of the Atlantic where the debris had plunged. By the time the Rogers Commission was created, efforts to recover and study the wreckage were already under way.

Within one day of the accident, NASA asked the National Transportation Safety Board to step in and assist with the investigation. The NTSB was experienced at piecing together clues after aircraft accidents, and despite the fact that *Challenger* was a spaceship, it was also an aircraft. It was hoped that NTSB staff members could tell what happened during the explosion, as opposed to damage caused by impact with the water. Some of the shuttle's parts rained down for as much as fifty minutes following the accident.

Two months of searching and waiting followed the *Challenger* disaster, until the day when the world would see for itself what was left of the shuttle. Reporters were called to the KSC Press Site to meet with the NTSB's Terry Armentrout, director of the effort to study the wreckage. The debris recovered up to that point was being arranged in a hangar, not far from the launchpad where *Challenger* lifted off for the last time. The day began in the Press Site auditorium with reporters firing questions at Armentrout. After a while, his voice betrayed his loss of patience. "I can paint word pictures all day, but you need to see it for yourself," he said.

Buses outside the Press Site were poised so reporters and cameramen could do just that. The caravan slowly threaded its way to the hangar. *Challenger*, which had disappeared two months before, was about to reappear.

The first sight that greeted the reporters was perhaps the most poignant. Outside the aircraft hangar, a flatbed truck had just pulled up. The shuttle's tail fin was lying on its side, held tight in place with cargo straps. Dive teams had just found and recovered it. The rudder itself retained its white appearance, accented with black heat protection tiles. The base of the fin, however, clearly showed the trauma of the accident. Torn metal and shredded cabling dangled from where the rudder had once been attached to the back of *Challenger*. On launch day, just below the rudder on the shuttle were the three main engines that helped propel the spacecraft off the pad. On either side of the fin were the two bell-shaped orbital maneuvering jets—the "big boomers," as NASA called them. Their job was to gently push the shuttle out of orbit at mission's end to begin the fiery descent from space to a safe landing. Now all these parts were just words on a list for salvage teams to find.

After surveying the rudder, the group moved inside. The hangar smelled of fish.

The floor was marked off with yellow tape forming a grid of four-foot squares. Straight ahead in the center of the room, the words *United States* glared out. It was a long jagged fragment of the shuttle's hull propped up on sawhorses. Specifically, it was a piece of *Challenger*'s midbody, halfway between the nose and the tail. Below this fragment was where the right-hand wing had once been attached. Above it was the spot where one of the clamshell payload bay doors once enclosed the shuttle's payload in the cargo bay. The doors were stacked nearby.

Terry Armentrout explained that the fish-market smell was due to the fact that much of the *Challenger* wreckage had sat on the bottom of the ocean for weeks, allowing barnacles to collect. Stench aside, this fact complicated the investigation. Chemicals were needed to remove the marine life that attached itself to the debris. But that treatment, it was feared, could erase evidence of the accident.[3]

At this point, some two months after the shuttle tragedy, about 20 percent of the orbiter had been found, either floating on the Atlantic or on the sea bottom. For the NTSB, each bit was a puzzle piece. The *Challenger* accident would pioneer many of the investigative tools that would be used following the *Columbia* disaster in 2003. Most notable was using the heat protection tiles to pinpoint where a certain piece of wreckage came from. The magic is in the numbers. When NTSB investigators gather wreckage from a commercial jet crash, the toughest part is telling one twisted piece of metal from another. Early in the *Challenger* investigation, NASA suggested using the serial numbers on the tiles that were still readable.

The shuttle's hull isn't a perfectly flat surface. The tiles need to hug the wings and curve around the bulbous pods that house some of the jet thrusters. That means each tile is custom cut to fit a specific spot. The numbers on the side tell technicians where a tile is to be glued on in a process called lick-and-stick. Those numbers are thus a roadmap to what might appear to be an anonymous piece of shuttle wreckage. Investigators comparing the wreckage to a schematic of the tiles have a better chance to solve a puzzle where all the pieces are white.[4]

Despite that one similarity, NASA's two shuttle accidents were vastly different. *Challenger* exploded just over a minute after liftoff, the shuttle was engulfed in a dramatic ball of fire, the boosters careened off in two directions, and bits of the spacecraft were seen flying from the flames. Investigators say the heat protection tiles did their job: they insulated the shuttle from the burning fuel from the external tank. That means the pieces appeared torn up but unburned.

By contrast, *Columbia* broke apart in 2003 while the shuttle was experiencing the worst of the fiery reentry into the earth's atmosphere. The superhot gases, at 3,000 degrees, left the *Columbia* wreckage charred.

Figure 19. *Columbia* wreckage: reaction control system thrusters, or RCS jets (cf. figure 13). These thrusters would have been on the right side of the vehicle beneath the flight deck windows where astronaut Willie McCool sat strapped in for reentry. Photo by Pat Duggins.

Gray was the common color of the *Columbia* debris that was laid out in a hangar of its own.[5]

In March of 1986, during the *Challenger* investigation, Armentrout's earliest comments were among the most chilling. Next to the hull fragment that read "United States" was the forward end of the underside of *Challenger's* nose. It was the part that cupped the two-story crew cabin where the astronauts spent their final moments. To Armentrout's experienced eye, the trick was to tell whether any given damage was caused by the explosion, by parts like the external fuel tank and booster rockets smashing into each other, or by the pieces hitting the water. The nose fuselage splashed hard into the Atlantic, but the aluminum structure wasn't bent out of shape. "There is evidence in this outer shell, the damage that we see, from past experience, there was some mass inside it," said Armentrout. "It wasn't like a piece of the external tank that just fluttered down." That was as close as the NTSB would come to saying that the crew cabin,

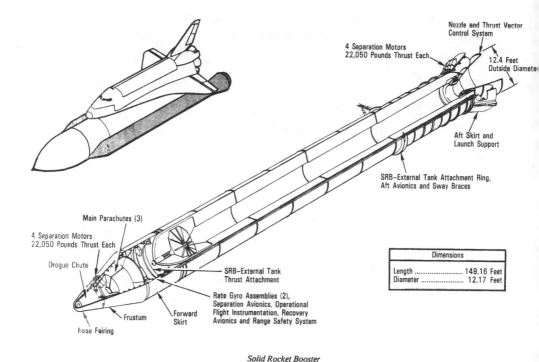

Solid Rocket Booster

Figure 20. Solid rocket booster. The shuttle's boosters generate 80 percent of the thrust needed to propel the spacecraft into orbit. The *Challenger* accident pointed up flaws in the design of the field joint that connects the metallic segments. The joint was redesigned to contain the hot booster gases and avoid the blowtorch-style leak that was blamed for the loss of *Challenger*. Courtesy of NASA.

with the astronauts inside, fell in one piece and smashed onto the ocean's surface.[6]

This evidence all made its way to the Rogers Commission for consideration. Like an onion that sat on the shelf too long, the more the panel peeled away the layers of data on NASA's solid rocket booster program, the less the board members liked what they saw. In the years prior to *Challenger*, NASA considered the solid rocket boosters one of the simpler components of the shuttle system, and among the least of the agency's concerns. The SRBs would go on to become one of the most controversial parts of these early flights.

In the early 1970s, Morton Thiokol Corporation in Utah competed against Lockheed, United Technologies, and Aerojet for the contract to build the shuttle's boosters, which would provide most of the thrust to put each spacecraft into orbit.[7] The Rogers Commission learned that when Thiokol won the deal in 1973, its bid failed to beat any of its competitors. The company did, however, offer a design that NASA judged to be more economical over the life of the contract, and that was enough.[8]

The boosters Morton Thiokol was hired to build were 150 feet long and 12 feet wide. They would be used in pairs on either side of the big external fuel tank with the shuttle bolted on. Their function was to provide enough thrust to carry the shuttle to an altitude of about 200,000 feet before being jettisoned. Parachutes would later pop from the nose cones so the empty boosters would fall gently into the ocean for recovery. Undamaged parts would be used again on later launches.

The rocket casings were hollow and filled with solid fuel that roughly resembled a big pencil eraser. The design was similar to boosters used on the unmanned Titan 3 rocket, since there wasn't just one long rocket casing but rather segments that were stacked one atop the other to form the long structure.

The can-shaped segments fit together with a tab-and-slot system called the "clevis and tang." The tang is the tab at the bottom of one segment that fits into the U-shaped clevis at the top of its lower neighbor. The two-part "field joint" also used rubber gaskets called O-rings to contain the booster thrust, as hot as 6,000 degrees, and focus it down through the rocket nozzle during its two-minute flight. The inside of each segment joint was coated with asbestos putty to help reinforce the seal.[9]

That was the way the rockets were supposed to work. Their actual performance was disturbingly different. The Rogers Commission found that the Thiokol boosters used looser-fitting clevis-and-tang field joints than the Titan 3 boosters that inspired the shuttle rockets, so one segment didn't nestle as tightly against the next. The unmanned Titan rocket used its boosters only once, while the parts in the shuttle boosters were meant to be used as many as twenty times.[10]

As early as the mid-1970s, concern was raised over rocket thrust leaking between the segments. Designers had hoped that under the pressure of a booster in flight, the clevis and tang would be pushed against each other to provide a strong seal. Unfortunately for the fledgling shuttle program, the opposite occurred. The parts opened up, creating a bigger gap in a phenomenon called "joint rotation." That would put even more reliance on the rubber O-rings to contain the hot booster thrust, something they were never designed to do. Thiokol's answer to the joint rotation problem was to use thicker O-rings, and that satisfied NASA.[11]

Witnesses told the Rogers Commission that the seal question persisted following the maiden launch of *Columbia* in 1981. The first incident of O-rings being burned by leaking rocket thrust during liftoff occurred during STS-2, the second launch of the shuttle. Inexplicably, word of the O-ring problem on STS-2 wasn't passed along to managers conducting the pre-launch reviews for STS-3, the next mission to go.

Testimony before the commission illustrated the degree to which officials at NASA debated the threat posed by the booster joints. It all came down to two designations, "criticality-1" and "criticality-1R." If the failure of a single component during a shuttle mission could cause the destruction of the vehicle and the loss of the crew, NASA called that component "criticality-1." If there was a redundant part that acted as a backup to save the spacecraft from the failure of the primary component, that system was called "criticality-1R." The field joints on the shuttle boosters were considered "criticality-1R" because there were two O-rings in each joint, which NASA felt would prevent extra protection. But then the boosters were tested under flight pressure, and the joint rotation demonstrated that the second O-ring didn't provide much help. Only then did NASA change the field joints' designation to the most vital "criticality-1"—in December of 1982, after twelve astronauts had flown on five missions.[12]

The problem of booster gases leaking past the asbestos putty and coming in contact with the rubber O-rings persisted. Commission members heard testimony that NASA became alarmed in 1983 and asked Morton Thiokol to change the putty used inside the rockets. The booster builder was later criticized in NASA memos for failing to act promptly on the situation. But rather than grounding the shuttle, NASA kept on launching, which led to three groups of consecutive shuttle launches with booster joint damage.[13]

The commission's final report states that, in 1984, two separate missions featuring *Challenger* and the maiden launch of shuttle *Discovery* had problems between booster segments or at the nozzle at the bottom. In 1985, four more *Discovery* missions and two additional *Challenger* flights suffered anomalies. Then, as NASA approached 1986, two *Atlantis* launches, one of *Columbia*, and one *Challenger* mission before the fatal

STS-51L accident encountered booster problems. In all of these incidents, there was either a gap in the asbestos putty that lined the inside of a field joint or burning of an O-ring.

In addition to concerns over the impact of hot booster gases on the joints, cold weather had also clearly become a factor.

A flight featuring shuttle *Discovery* in January of 1985 included the coldest pad temperatures at launch, at 53 degrees. That shuttle suffered not only erosion of the asbestos putty but also a complete failure of an O-ring where hot thrust gases leaked through. The astronauts survived that 1985 incident. A year later, freezing temperatures would prove fatal for *Challenger*'s final launch.

When NASA found yet another damaged O-ring following a shuttle flight in April of 1985, agency managers ordered that the booster joints be listed as a "constraint to launch." That meant that launch controllers had to resolve the problem or sign a waiver to continue with the countdown. NASA proceeded to sign waivers on six missions until *Challenger* was destroyed in late January of 1986.[14]

Along with analysis of the wreckage, launch video of the shuttle disaster uncovered the "smoking gun" almost immediately. At booster ignition, enhanced images showed several puffs of brown and black smoke coming from the suspect field joint of the right-hand booster rocket. NASA engineers theorized this was the asbestos putty and then the O-ring burning through.[15] Then, as the shuttle soared off the pad, a plume of white-hot booster exhaust was seen streaming from the side of the right-hand booster rocket. That gave salvage crews a specific idea of what parts to look for in the Atlantic, but help was needed to hoist the huge pieces of wreckage off the ocean bottom. NASA called in the vessel named *Stenna Workhorse*, which had cranes powerful enough to bring debris and evidence of the *Challenger* accident to the surface.[16] The easiest task was finding wreckage that floated. Parts of the external fuel tank, the tank's foam insulation, and the heat protection tiles that covered the shuttle like a mosaic fluttered down to the Atlantic and stayed on the surface. More critical pieces like the booster that failed, the main engines, and the crew cabin with the astronauts perhaps still inside sank to the bottom of the sea.[17]

All of this data and testimony went before the members of the Rogers Commission to establish a long history of problems with the boosters.

When NASA prepared to launch space shuttle *Challenger* on the ill-fated STS-51L flight, the panel would learn, miscommunication between Thiokol engineers and NASA managers compounded the inherent flaws in the boosters to create tragic results.

Challenger's first launch attempt, on January 27, 1986, was scrubbed because of unacceptable crosswinds high over the pad. The weather forecast for the next day called for temperatures as low as 27 degrees, the coldest ever for a shuttle launch. As a result, top NASA managers met with the agency's own engineers and with Morton Thiokol personnel, in teleconference, to discuss what might happen to *Challenger's* boosters in the freezing weather. The meeting had been prompted, in part, by the near disaster in 1985 involving shuttle *Discovery* when O-ring erosion occurred during comparatively warmer temperatures of 53 degrees. One man summoned to testify before the Rogers Commission to describe the meeting was Roger Boisjoly, a Morton Thiokol engineer working on solid rocket motor seals. He had participated in the crucial exchange, and was part of a vocal minority that was extremely concerned over the weather issue.

Boisjoly testified that he was asked to make his case to NASA on what might go wrong with the boosters if the countdown proceeded. "I was asked to quantify my concerns, and I couldn't," said Boisjoly. "I had no data to quantify it, but I knew it was away from the goodness in the data base." By that, the booster specialist meant that the rockets hadn't been tested and certified to withstand that type of cold. Boisjoly said he continued to press the issue of the weather-related O-ring erosion on shuttle *Discovery* the year before, and how inspection of the damaged booster rocket indicated clearly that the ambient temperature was a contributing factor. Commission chairman William Rogers asked what the conclusion of his argument was, and Boisjoly said that NASA should not launch the shuttle when it was colder than the 53-degree mark that *Discovery* endured in 1985. "We were quite pleased because we knew in advance, having participated in the preparations, what the conclusions were," said Boisjoly. "We felt very comfortable in that."[18]

Not everyone shared that view. Another participant in the prelaunch teleconference was Lawrence Mulloy, manager of NASA's solid rocket booster program. When presented with the weather concerns of the Thiokol engineers, Mulloy's response became one of the most popular quotes in the press during the *Challenger* investigation. "My God, Thiokol, when do you want to launch, April?" he was reported to have said. Mulloy was asked to testify about the Thiokol teleconference, not just the much-publicized exclamation. The NASA manager said his statement was taken out of context. "I have never heard anyone who says they heard those words," said Mulloy. Rather, he claimed to be questioning what he believed to be Thiokol's last-minute attempt to rewrite the launch commit criteria, which is a list of conditions NASA must adhere to in order to keep the countdown clock ticking to zero. "If we do not launch without a temperature greater than 53 degrees, we may not be able to launch before April," said Mulloy of the discussion. "We need to consider this closely before we jump to any conclusions," he recalled adding during the teleconference.[19]

After much debate, including a private caucus by Thiokol officials, the booster builder decided that the O-ring concerns weren't reason enough to cancel the launch. The result flew in the face of Roger Boisjoly's recollection of the exchange over *Challenger* and its boosters. Chairman Rogers asked the witness if Thiokol's recommendation not to launch the space shuttle under the current weather conditions was unanimous or not. "I've been distressed about the things appearing in the paper and things that have been said in general," replied Boisjoly. "There was never one positive prolaunch statement made by anybody." The commission concluded that Thiokol's decision that the data surrounding the O-ring issue was inconclusive, and that NASA should launch the space shuttle, happened at the urging of midlevel NASA managers.[20]

The arguments brought forth by Roger Boisjoly and other Thiokol engineers made it no farther up the chain of command at NASA. Neither the Kennedy Space Center director nor the shuttle program manager nor the launch director said they had heard anything about it. This disturbed commission member General Donald Kutyna, who was director of Space Systems and Command for the U.S. Air Force. He inquired of Mulloy why the decision to go ahead and launch had apparently been made by NASA managers at the seniority level of those involved in the teleconference, and not by people farther up the decision chain. Mulloy explained that his supervisor, shuttle program manager Stanley Reinartz, who had taken part in the meeting with Thiokol, was part of the launch team decision-making group. "If this were an airplane, an airliner," said Kutyna, "and I just had a two-hour discussion with Boeing on whether the wing was going to fall off or not, I think I would at least tell the pilot, or at least mention it." To what levels of NASA management, the general asked Mulloy, did the information from the Thiokol teleconference go? "As I stated earlier," Mulloy responded, "Mr. Reinartz, who is my manager, was at the meeting, and on the morning, about five o'clock, in the operations support room where we all were, I informed Dr. Lucas of the content of the discussion." William Lucas was director of NASA's Marshall Space Center in Alabama, which was the installation in charge of the shuttle's engines and boosters. The commission asked how high that kind of revelation would normally go, and the answer was, to the shuttle program manager and the associate administrator of spaceflight. The agency admitted there was a serious breakdown in communication. This helped confirm that there was more wrong in the shuttle program than a faulty booster design.[21]

The Rogers Commission decided that a redesign of the booster was needed. The new rocket would incorporate a modified field joint with a third O-ring to enhance safety. NASA was also sharply criticized for a flawed management system in which the initial concerns over the booster joints were prevented from reaching top decision makers. Launch managers were also taken to task for signing waivers to keep the shuttles launching despite ongoing concerns. Morton Thiokol was also singled out for failing to press its case over the cold weather's possible impact on the O-rings in order to keep a big customer happy. The final report didn't satisfy some critics, who expected the document to address speculation in the press that top managers at NASA played a role in the decision to launch, and that the schedule for President Reagan's State of the Union address before Congress in January of 1986 was a factor—a point vehemently denied by the White House during and after the investigation.[22]

While NASA took its lumps, reworked its corporate culture, and pressed ahead with remaking its solid rocket boosters, the media persisted with questions on what the *Challenger* astronauts experienced during the explosion and later, before their crew cabin hit the water. Former astronaut Richard Truly, now a top NASA manager, asked physician and former Skylab crew member Joseph Kerwin to examine data from the accident with an eye to answering that question.

In a letter dated July 28, 1986, Kerwin delivered his findings. Despite the fireball that engulfed *Challenger* and caused the vehicle to break up, the blast was judged insufficient to kill the crew. Kerwin determined that the blast likely blew the crew cabin free of the shuttle, causing an initial

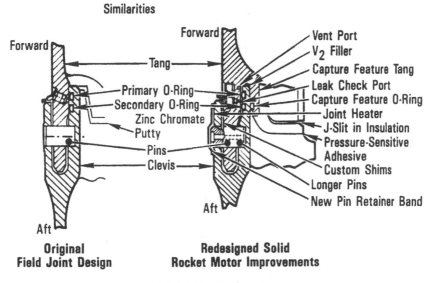

Similarities

Forward

Tang

Primary O-Ring
Secondary O-Ring
Zinc Chromate
Putty

Pins

Clevis

Aft

Forward

Vent Port
V$_2$ Filler
Capture Feature Tang
Leak Check Port
Capture Feature O-Ring
Joint Heater
J-Slit in Insulation
Pressure-Sensitive
Adhesive
Custom Shims
Longer Pins
New Pin Retainer Band

Aft

**Original
Field Joint Design**

**Redesigned Solid
Rocket Motor Improvements**

Field Joint Comparison

Figure 21. Comparison of the older field joint on the space shuttle's solid rocket boosters (left) with the redesigned field joint following the *Challenger* accident (right). An extra O-ring was added to help prevent white-hot booster gases from leaking out as they did on *Challenger*. That problem was blamed for NASA's first shuttle disaster, which killed seven astronauts, including teacher-in-space Christa McAuliffe. Courtesy of NASA.

increase in acceleration as the compartment sailed upward for twenty-five seconds and then went into a nosedive toward the Atlantic. The cabin hit the water at a speed of more than 200 miles an hour almost three minutes after the initial explosion. Kerwin judged that the impact with the water would not have been survivable in any event. The condition of the straps and buckles that held the astronauts in their seats led Kerwin to believe that the crew remained firmly in their flight couches during the trip down.[23]

Perhaps the most compelling evidence that the astronauts were at least somewhat aware of what was happening to them was to be found in their "personal egress air packs" or PEAPs. Each crew member carried an emergency air supply in a unit shaped like an attaché case. In the event of a problem during liftoff, they could flip a switch and receive air through a tube going from the PEAP to their space helmets.

Kerwin says there was evidence that at least three of the units had been activated. The possibility of the switches being thrown accidentally by the force of the explosion or splashdown in the Atlantic was ruled out.[24]

NASA Falls Short on Its Promise to the Air Force

As NASA dealt with the pain of losing the Challenger Seven and worked to resume shuttle flights, it became obvious to people like former secretary of the air force Robert Seamans that promises made to persuade the military to endorse the shuttle wouldn't be fulfilled. The bargain appeared unlikely to be met even if the program flew again successfully. Prior to *Challenger*, expensive modifications were made at Vandenberg Air Force Base to accommodate California launches for the Pentagon, but no West Coast blastoffs ever occurred.

That's not to say NASA didn't do its share of hush-hush missions, the first of them in 1985. There would be nine secret launches where neither NASA nor the astronauts talked about the payloads or their objectives. Launch controllers even kept the exact countdown quiet, supposedly to keep inquisitive eyes from foreign powers from knowing where to look as a certain shuttle sailed to orbit.[25] The secret missions did afford the media a bit of amusement. One reporter was seen walking around before a blastoff wearing a T-shirt emblazoned with the plans for a satellite known by the acronym DISCUS. That was the classified payload thought to be aboard the shuttle, though NASA wasn't saying.

The Pentagon was happy enough with its deal with NASA to assign an astronaut crew to make the first shuttle launch out of Vandenberg in 1986. The flight was called STS-62A and it would have teamed astronauts Bob Crippen, Guy Gardner, Jerry Ross, Mike Mullane, and Dale Gardner with air force crew members Edward Aldredge and John Watterson. Then, with the *Challenger* disaster, everything changed and the first Vandenberg mission was cancelled.[26]

So for all the effort and concessions that led to the birth of the space shuttle, the air force got only a fraction of the missions it was promised. The lack of even one Vandenberg liftoff angered former air force secretary Robert Seamans. The Pentagon had pressed ahead with this partnership with NASA despite his misgivings, and now the situation had turned around to bite the military. "We put a few billion dollars into that [California] launch site," says Seamans. "And all that was wasted. I think from the air force's standpoint, it was unsatisfactory."

Once NASA resumed shuttle launches in 1988, its lack of a defined mission would become even more obvious. The cancellation of any military shuttle launches from Vandenberg meant NASA's cloak-and-dagger days were numbered. In August of 1986, President Reagan also ordered NASA not to accept any new contracts to launch commercial satellites from the shuttle.[27] The White House advocated the creation of a U.S. space station called Freedom, but it would be another decade before the first U.S. component was launched. Without military payloads or commercial satellites to carry, NASA lost two reasons for the creation of the shuttle program. And there was still no mission.

1988–1992

Back on the Horse

Despite a dwindling number of classified military flights and no commercial satellites to carry as cargo, NASA appeared to have a lot on its plate as it struggled to revive the shuttle program between 1986 and 1988. There still wasn't a central purpose for the shuttle, other than to fly again in honor of the lost *Challenger* crew. But that lack of a future wasn't apparent to the public during the first return-to-flight missions. There was a backlog of NASA science observatories and interplanetary spacecraft to be carried to orbit on the shuttle, as well as the agency's Tracking and Data Relay Satellite System to deploy. The TDRSS satellites were designed to funnel pictures and computer data from orbiting spacecraft to ground controllers and scientists.

Before any of that could happen, NASA had to prove that the shuttle was safe to fly again, and that the agency had regained some measure of credibility following the embarrassing revelations of mismanagement made by the Rogers Commission. One group of astronauts would find themselves in the limelight during this time. Before *Challenger*, veteran shuttle crew members Rick Hauck, John Lounge, and Dave Hilmers were training to carry and launch a robotic European space probe called Ulysses on a mission to study the sun. Those plans changed one day without fanfare. "They [NASA] didn't name the reflight crew until a year after the accident," Lounge says. "I was training with Rick Hauck and Dave Hilmers, and they just kind of moved us into that slot." The trio were now assigned to fly the first shuttle to go after *Challenger*, in 1988.[1]

The Ulysses probe would launch on a later shuttle mission, in 1990. The composition of its crew, all of whom made their rookie flights after the *Challenger* accident, served as a sign that NASA was moving on. The shuttle that would accompany the solar probe to orbit also utilized the redesigned solid rocket booster. This was supposed to make the space shuttle safer than ever, but the first men to strap themselves on board and try the booster out didn't find it any easier. "The older I get, the more I know how fearful I was on the launchpad," says *Discovery* astronaut Dick Covey.[2]

Covey, Hauck, Hilmers, and Lounge, were joined on the first post-*Challenger* crew by fellow veteran George "Pinky" Nelson. Covey and Lounge had flown together in 1985 on a satellite retrieval mission. Hauck commanded *Discovery* on a satellite rescue flight of his own. Pinky Nelson performed a spacewalk in 1984 to try to capture a wayward satellite, and Hilmers had been on the maiden flight of shuttle *Atlantis*. Like *Apollo 11*, the first shuttle mission following *Challenger* had no rookie astronauts. The flight plan was equally conservative. It was a bare-bones mission featuring the deployment of a Tracking and Data Relay Satellite known as TDRS-C. Its job was to replace TDRS-B, which was destroyed in the *Challenger* accident. Other reminders of *Challenger* would follow. The press kit for STS-26 was emblazoned with the phrase "NASA's Return to Spaceflight!" News articles and broadcast reports focused on *Challenger* and the lost astronauts and how NASA's future hinged on the success of the *Discovery* mission. "Liftoff was worth it, just to get away from the [media] fishbowl," says John Lounge.[3]

That's not to say that the first crew to fly on a shuttle since the *Challenger* accident didn't have their share of butterflies in their stomachs. The first minute or so was especially tense, since *Challenger* exploded just after the spacecraft passed a point in flight called Max Q, or maximum aerodynamic stress on the vehicle. To minimize the impact on the shuttle, the craft is programmed to throttle its engines down for a bit and then rev them back up as the air starts to thin out. That point is marked by a traditional call from Mission Control in Houston. The capsule communicator there is the one person who talks to the astronauts by radio. The capcom says, "Go at throttle up," and the shuttle commander responds usually with "Roger, go at throttle up." That was the last thing *Challenger* commander Dick Scobee said before the shuttle exploded, and the phrase became engrained in the public's collective memory as the cue for the *Challenger* tragedy. During the launch of shuttle *Discovery* on the first mission after the accident, commander Rick Hauck departed from tradition during the trip to orbit.

"*Discovery*, go at throttle up," capcom radioed.

"Roger, go!" Hauck responded.[4]

The trip from the launchpad to earth orbit usually takes about eight minutes. After that, the main engines shut down and the crew sometimes prepare for a thruster firing to fine-tune their path around the globe. *Discovery* astronaut John Lounge recalls how agonizingly slow the trip to orbit was on STS-26. His first flight had been aboard *Discovery* with crewmate Dick Covey in 1985. Back then, the eight-minute trip zipped by. To Lounge it felt like four minutes instead. The first post-*Challenger* flight was a lot different. "Each Mach number [on the speedometer] was just creeping by," remembers Lounge. "It was like sixteen minutes." *Discovery*'s mission went according to plan, however, and the craft landed safely.

For a time following STS-26, scientists and the Pentagon clamored to have their payloads carried into orbit. Perhaps the most critical task was the rescue of the 20,000-pound Long Duration Exposure Facility, or LDEF. The unmanned orbiting satellite contained fifty-seven experiment trays to test, among other things, metallic coatings that could be used on future spacecraft. Other experiments included plant seeds that scientists wanted to expose to the conditions of space. LDEF was stranded in space by the *Challenger* accident, and its orbit was rapidly decaying. Much longer and it would burn up as it plunged into the earth's atmosphere. Astronaut Dan Brandenstein would lead a crew aboard *Columbia* to retrieve it.

The astronauts would also have to get used to changes prompted by the *Challenger* accident. Along with modified solid fuel booster rockets, the astronauts faced an altered routine. The blue cloth flight coveralls that shuttle crews had worn for launch and landing since STS-5 were ruled out. All the astronauts would have to wear a launch-and-entry suit, or LES. The flight suits include parachutes and survival gear, and each weighs close to ninety pounds regardless of the stature of the astronaut. Besides the added weight of the LES, there were bulky helmets and gloves to deal with while flying the shuttle. Astronauts found their reach restricted and their peripheral vision limited.[5]

"Most important with the darned things," says Brandenstein, "is that the commander goes in first. So you spend a lot of time on your back [in a flight seat]. And, the orange suits were a lot more uncomfortable than the plain cotton suits."

Programming was added to the shuttle so the astronauts had a bailout option in case of emergency. That procedure meant putting the shuttle into an autopilot glide, opening the main hatch with explosive bolts, and extending a long metal pole out and away from the shuttle. The astronauts would each clip to the pole a ring attached to their flight suits, slide down the pole, and then open their parachutes once they were clear of the shuttle.[6] Critics complained that the bailout plan would work only if the shuttle was in a perfectly controlled glide and only during the final stages of the trip back to Earth.[7]

Dan Brandenstein and his STS-32 crew pressed ahead with their mission to rescue the Long Duration Exposure Facility despite its painful connections to the *Challenger* disaster. LDEF was first deployed in space during a *Challenger* mission in mid-1984. *Columbia's* crew, including rookie astronauts Jim Wetherbee, Marsha Ivins, and David Low and veteran Bonnie Dunbar, launched the Syncom-V satellite for the Defense Department, then set their sights on finding LDEF. Part of their job was to document the condition of the satellite before trying to grab it with *Columbia's* fifty-foot robotic arm. "It looked like it had more damage than we were anticipating," observes Brandenstein. "Some of the foil coverings had curled up. We expected it to have a lot of micrometeorite dings. But it looked almost like those cartoon tuna cans with the top curled back." After *Columbia* landed, sections of LDEF were put on display for members of the press to

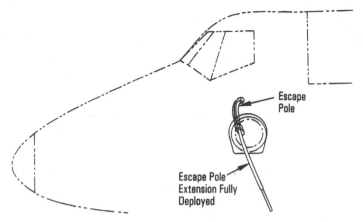

Side View of Escape Pole Deployed Through Side Hatch

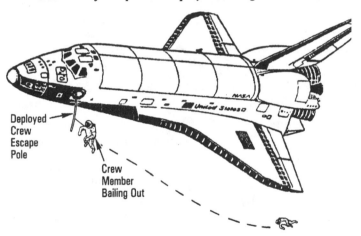

Crew Bailout Mode—Crew Escape Pole

Figure 22. Bailout option. A system for escape from the space shuttle was instituted following the *Challenger* accident. The crew must put the shuttle into a controlled glide, blow the hatch open, and extend a long pole. Then each astronaut slides down the pole and parachutes into the ocean for later rescue. Courtesy of NASA.

see. Some of the smallest impacts on the metal frame left scars that looked like tiny comets. A micrometeorite's initial hit would create a minicrater, with a skid mark tailing off.[8]

NASA Builds a Replacement Shuttle
to Fill the Void Left by Challenger

NASA signed a $400 million contract during the construction of *Discovery* and *Atlantis* to create a set of spare parts. An extra set of wings, a rudder and speed brakes, engine compartment, midfuselage, and components for the nose section were manufactured and set aside.[9]

NASA would need them now.

The Astronaut Office also needed someone to act as liaison as the new space shuttle was being built. The job went to rookie Bruce Melnick, a member of the first astronaut class to be selected after *Challenger*. Melnick had applied for every astronaut class since the first group of shuttle candidates were picked in 1978. NASA was doing his background check as investigators searched for the cause of the shuttle disaster.

Melnick's first assignment was with a small group of astronaut candidates known as Cape Crusaders, whose job it was to oversee work at the Kennedy Space Center. When one of them was needed to supervise construction of the replacement shuttle, Melnick says, "I lucked into the job because I was a pilot. I could fly a T-38 training jet, and I was the only one who could do that." Melnick had been a test pilot who flew high-performance helicopters for the Coast Guard. His job then was to fly the new HH38 Dolphin copter at speeds of up to 165 miles per hour and make landing attempts on search-and-rescue vessels pitching and rolling in stormy seas at night. The goal was to push the aircraft to its design limits. His pilot's license was used a lot while observing the construction of NASA's newest shuttle. "Monday, I would fly myself to Kennedy Space Center and attend meetings there," he recalls. "Then Tuesday, I'd fly to Palmdale and talk to people working on the shuttle there. Then back to Houston on Thursday." Melnick kept up that pace for months. He was still waiting on his first flight in space, but seeing a shuttle from the inside out during the new orbiter's construction gave him a unique appreciation of the spacecraft and a unique distaste for the heat protection tiles. "It's just a pretty-designed aircraft," says Melnick. "Beautiful curves, just like an airplane, and then you put all those heat tiles on it and it looks like a slab."

Work to build the new space plane actually began in 1982 before the maiden flight of *Challenger*. That's when NASA's prime shuttle contractor built a crew compartment as a test article. As the new craft came together, Melnick found time to joke with the engineers at Rockwell who were doing the assembly work. "I always told them that I was going to be on the maiden flight of that shuttle," he says.

The astronaut flew back to Houston after one of his round-robin trips between the Johnson Space Center, KSC, and California, where *Endeavour* was being built. The pledge he made to Rockwell engineers that he would have a seat on the first flight of the new shuttle was still fresh in his mind when the phone rang. It was the home office of the astronaut corps. Melnick had his rookie assignment on a space mission. The catch was that it wasn't aboard *Endeavour*.

Shuttle Discovery Carries a Satellite to Study the Sun

Following the *Challenger* accident, there was a backlog of interplanetary missions that needed to go, and Melnick was being assigned to a flight aboard shuttle *Discovery*. It was to launch the European solar probe called Ulysses. Melnick had never heard of it.

"Couldn't even spell it," he says.

Ulysses was stuck on the ground following *Challenger*, along with the Galileo spacecraft bound for Jupiter, the Magellan mission to Venus, the Hubble Space Telescope, two NASA Tracking and Data Relay communications satellites, and a number of hush-hush missions for the Pentagon.

Ulysses was a scaled-back version of what once was a joint mission between the United States and Europe called the International Solar Polar Mission. Its point was to fly over the poles of the sun in a never-before-tried path called "out of the ecliptic." One thing many interplanetary satellites have in common is where they fly. It's on the same two-dimensional plane as the course the planets follow around the sun. To do what its European designers wanted, Ulysses would have to whip around the planet Jupiter to turn that flat path on end, so the 800-pound probe would circle the sun's north and south poles, instead of going round and round its equator like the planets.[10] There were no cameras on Ulysses to send back dramatic pictures to hang on the wall.[11] But Melnick didn't care. He got his shuttle mission, which he considered a great stroke of luck because he felt he bumped more experienced people to get it.

Those who are picked to be astronauts spend their first year in astronaut candidate training—basic instruction in survival gear, astronomy, meteorology, and so on. Year number two is spent in what's called pilot pool training. That's where shuttle pilots start to fly training jets and the remaining rookies, the mission specialists, learn how to operate the robotic arm and make spacewalks. Melnick was picked in 1987 as an astronaut candidate, and his initial training ended in 1989. After that, he was put in line behind people from two previous classes to wait for a mission. "I was blown away," says Melnick of his selection for the Ulysses mission. "You didn't care what payload you got as long as you got a mission." Now Melnick and his crew, commander Dick Richards, pilot Bob Cabana, and fellow mission specialists Tom Akers and Bill Shepherd, got in line behind other flights that were grounded because of the *Challenger* accident. Melnick was convinced that would mean more waiting as shuttle after shuttle blasted off over the months to come.

Then luck came into play again during what veterans at NASA refer to as "hydrogen summer."[12]

Columbia was on the launchpad in May of 1990 for its third mission since the shuttle disaster when it developed a dangerous hydrogen leak around its engine compartment. Specifically, fuel was seeping from the seventeen-inch-wide pipe that feeds explosive propellant from the big orange external fuel tank to the vehicle's three liquid-fueled main engines. The shuttle fleet was grounded for six months while engineers worked to fix the problem. But while this was going on, the clock was ticking for Ulysses as well.[13]

The mission called for launching the space probe on a path toward Jupiter so it could use the giant planet's gravity to whip around into its top-to-bottom path around the sun's poles.

Timing was everything.

Hydrogen concerns or not, space shuttle *Discovery* had to blast off within an eighteen-day period in October of 1990; otherwise NASA and the Europeans would have to wait another thirteen months for the earth and Jupiter to line up again. *Discovery* wasn't leaking, so NASA chose to proceed with the launch of Melnick's mission, called STS-41.

Training for the mission gave the rookie astronaut a chance to get acquainted with his payload and to study up on solar astronomy. Ulysses had a ten-foot-wide sandwich-shaped box for a body, wrapped in gold foil. The structure was topped with a five-foot-wide main antenna to beam its data back to Earth. The jet-black dish was marked in gold with the letters ESA for the European Space Agency.[14]

Deploying Ulysses from *Discovery* meant dealing with a timeline even tighter than the launch window for the shuttle itself. The small space probe, perched on top of a thirty-foot-tall white and gold two-stage rocket nestled in the cargo bay, had to go during a launch opportunity lasting only a few seconds.

For Bruce Melnick and his STS-41 crewmates, sending Ulysses on its way would occur just six hours after blastoff. That was a time when the crew would feel at its worst during the five days of weightlessness. After the rough ride to orbit, the fluid inside the body floats up into the chest and head. That leaves the space traveler feeling stuffed up and frequently woozy from the onset of space motion sickness and possibly nausea. "You're not feeling great," says Melnick. "Your inner ear is shutting down, your digestive tract is shutting down. So, getting that thing [Ulysses] going was a great relief."

Once the astronauts were in orbit and the payload bay's clamshell doors were opened, the crew could look out onto Ulysses and its booster rocket. The next hours were spent leaning the rocket up on a tilt table until it was aimed at an angle to take Ulysses out over the crew cabin. "It had springs to push it away," Melnick remembers. "It went at a forty-five-degree angle over our heads and past the windows. It was spectacular!" The astronauts didn't have much chance to look out the windows and appreciate the view from space until after Ulysses began its yearlong trip to Jupiter and on to study the sun. Then the crew's attention turned to experiments on the shuttle's middeck, which filled their time until landing.[15]

The backlog of planetary probes and DOD missions soon dried up, and the post-*Challenger* shuttle program entered an era of science missions. There were Spacelab flights, there were excursions to examine the atmosphere, and there were spacewalks to practice repairing the Hubble Space Telescope, which had been launched with a flawed main mirror that blurred its vision.

Following *Discovery*'s return from launching the Ulysses probe to the sun, astronaut Bruce Melnick was assigned to the more routine job of testing avionics that would be used on later shuttle missions. That meant long hours in a laboratory in Houston testing navigation boxes under the conditions the gear would face during trips in orbit.

Then the phone rang again, with news that Melnick really would get to fly on the maiden flight of NASA's new shuttle that would replace *Challenger*.

Three Spacewalkers Reach Out to Save Intelsat 6

While construction of the new spacecraft dragged on, one element that still had to be settled was its name. In the late 1980s, the Bush White House took the more political route of leaving the name selection up to schoolchildren. Entries from Senatobia Middle School in Mississippi and Tallulah Falls School in Georgia were picked as the winners. Their choice was to call the new space shuttle *Endeavour*, the same name as the mother ship during the *Apollo 15* mission to the moon. The students' school colors can be found hidden in the embroidered crew patch for the maiden mission. The shuttle was named after HMS *Endeavour*, one of the ships that Captain James Cook used to explore the Great Barrier Reef and New Zealand. The sailing ship is depicted in the center of the STS-49 crew patch for *Endeavour*'s first flight. Flags with the school colors of Senatobia and Tallulah fly from the top of the ship's masts.[16]

Astronaut Dan Brandenstein would command the flight. Building *Endeavour*, says Brandenstein, "was important because it demonstrated that the country wasn't going to fold its tent following the [*Challenger*] accident. We thought it was kinda cool that the name was the result of kids working science projects and contests to pick the name." The downside for NASA was that *Endeavour* would be the nation's last shuttle. No more spares were available, and no new orbiters would be built, no matter what.

Endeavour's maiden flight, when Dan Brandenstein's path and Bruce Melnick's would cross, proved to be a major undertaking. What began as one group's misfortune ended as a memorable shuttle mission that NASA would use to remind its critics how much astronauts could accomplish in orbit.

The Intelsat Consortium had launched its most sophisticated satellite, Intelsat 6, on a Titan 3 rocket in 1990. An onboard malfunction prevented the separation of the Titan's second stage from the satellite and its small booster rocket. The booster was needed to push Intelsat 6 to its proper orbit, so paying customers could make use of the satellite's 120,000 phone channels. Intelsat 6 was also designed to funnel three television channels back to audiences on Earth. The technical problem forced launch controllers to command the satellite to drop the small booster along with the Titan upper rocket stage. That meant the two-ton Intelsat 6 was stuck in the wrong orbit, basically a $150 million paperweight.[17]

Intelsat asked NASA to stage a rescue. This would be the main task on *Endeavour*'s maiden flight, STS-49. The difficulty was that, unlike such satellites as the Hubble Space Telescope, Intelsat 6 was not designed to be nabbed by spacewalkers. There were no handrails for the astronauts to grab while wearing the heavy gloves of their spacesuits. Worse yet was the threat of Intelsat 6's sharp edges, which could puncture a suit and endanger the wearer. Regardless of the danger, it was a flight in which astronaut Bruce Melnick wanted to take part. "It was considered the mission everyone and his brother wanted to do," he recalls. "It was before we started building the International Space Station, so everybody wanted to go." Melnick's joke to Rockwell workers about making the maiden flight of NASA's new shuttle was about to take a couple of twists. Melnick had just got home from a long day at the avionics lab when Dan Brandenstein called to say he wanted Melnick on the *Endeavour* crew. Melnick's answer was emphatic. "I remember saying, 'Oh, man, yes! I'd love to go and do a spacewalk on that flight!'"

Brandenstein had a mixed bag of news on that account. Melnick would be assigned to the mission, but not as one of the spacewalkers who would chase Intelsat 6. His job would be the less glamorous one of robotic arm operator, and it was his experience as a Coast Guard search-and-rescue pilot that attracted Brandenstein's attention.

The flight plan called for astronauts Pierre Thuot and Richard Hieb to float outside on a spacewalk. Thuot would clamp his feet to the end of the fifty-foot robotic arm and, while holding a long metal pole called a capture bar, would be eased up close to Intelsat 6 so he could snap the bar to the bottom of the satellite. The robotic arm would use the capture bar's "grapple fixture" or handhold to grab Intelsat. The crew would then equip the satellite with a new booster rocket brought along in *Endeavour*'s payload bay so it could be launched on the mission its builders intended. The robot arm needed someone who worked with precision, and Brandenstein felt Bruce Melnick's background gave him an edge. Even so, the strategy wasn't fully endorsed by the astronauts. "I was never comfortable with the capture bar," says *Endeavour* commander Dan Brandenstein. "You had to click those latches with a satellite that's still spinning to keep it stable. My concern is that it wouldn't take much to send it wobbling." Besides Brandenstein, Melnick, Thuot, and Hieb, the crew would include Kevin Chilton, Kathy Thornton, and Tom Akers, who flew with Melnick on his rookie mission to deploy the Ulysses probe.

Endeavour's blastoff from the Kennedy Space Center took place on May 7, 1992, an exceptionally clear day. Conditions were so favorable, onlookers at the KSC Press Site could see the solid fuel boosters separate and drop away from the shuttle two minutes into the liftoff. The Intelsat consortium was there to cheer the astronauts on. Company representatives handed out plastic commemorative clip-on tags in bold yellow and black that were emblazoned with the slogan "From the Brink, Up to Sync." From the brink of disaster, that is, up to the proper geosynchronous orbit— assuming the *Endeavour* crew delivered a rescue as promised.

After two days in orbit, *Endeavour* rendezvoused with the seventeen-foot-long shiny black cylinder that was Intelsat 6. Crew members Thuot and Hieb donned their heavy pressure suits, squeezed into the shuttle's tiny airlock, and moved outside to try to capture the satellite.

Thuot found that Intelsat 6 was more easily bumped off kilter than training on the ground had indicated. His attempt to snap the capture bar in place sent the satellite into a drunken wobble. "The lowest I've felt during a mission was that first day when we were unsuccessful at capturing the satellite," says Brandenstein. Thuot and Hieb backed off while ground controllers tried to stabilize Intelsat with its onboard thrusters. The spacewalk was a failure.

The next day the spacewalkers ventured out again, but this time with some rehearsal in mind. Thuot and Hieb took the capture bar to the side of *Endeavour*'s payload bay, called the sill, to practice pushing the bar toward its target. With renewed confidence, Thuot snapped the feet of his spacesuit to the end of the arm and moved in closer to Intelsat. The extra rehearsal didn't help, and the bar didn't snap into place. "At that point, I knew the second day had gone as well as humanly possible," says Brandenstein. "I radioed Mission Control that everybody should take a day off and come up with a new idea." *Endeavour* fired its jets and moved to a safe distance from Intelsat 6, no closer to its quarry than before. It was the end of a long and disappointing day.

After dinner the astronauts were supposed to go to bed in the lower part of the crew cabin, the middeck. The radio chatter from Mission Control came to an end, and *Endeavour* was left to circle the earth in silence. But the quiet didn't last long. Dan Brandenstein wasn't in the mood to sleep. He left his crewmates and floated up to the upper part of the cabin, the flight deck, where *Endeavour*'s cockpit was located, as well as most of the windows opening out into the payload bay and toward Intelsat 6 in the distance. Before long Brandenstein was joined by pilot Kevin Chilton, who was just as restless. "When you miss a satellite two days in a row," says Brandenstein, "sleeping is the last thing on your mind. So, it was me and Chilton looking out the back window into the cargo bay to see what we had to work with."

The activity above didn't escape the notice of the astronauts in their sleeping bags on the middeck. "It was dark and I was trying to sleep," says Melnick. "Then I noticed a light go on up in the flight deck and heard talking. Rick Hieb was the first one where curiosity killed him, and he went up too."

In about fifteen minutes the whole crew had crowded onto the flight deck. The group began brainstorming on how to fix a problem that months of planning and simulations on Earth had failed to solve. The capture bar, they felt, had to go. The crew tossed around scenarios, the most radical of which involved having not two but three astronauts suit up and float outside to try to capture the satellite.

Bruce Melnick was in charge of the robot arm during spacewalks, but he also called out commands to the spacewalkers and kept track of their gear. Once the Intelsat issue was resolved, a different pair of spacewalkers was to go out and practice building techniques for the space station. "That meant we had three complete spacesuits," says Melnick. "I suggested sending three people out to make a three-legged stool." The idea was easy to suggest but tough to get past Mission Control. Each time spacewalkers venture outside the shuttle, it's a complicated and potentially dangerous process. Prior to a work session, the astronauts pre-breathe pure oxygen to purge nitrogen from their blood. Nitrogen bubbles in a spacewalker's bloodstream can be fatal, just as for a scuba diver who gets the "bends."

An airlock is located on the middeck of the spacecraft. The spacewalkers each don a 250-pound suit, which goes on like a shirt and pants that clamp together at the beltline, and snap on their helmets and gloves. The two astronauts float through a D-shaped hatch on the airlock, and the compartment is closed up and the air pumped out.[18] The spacewalkers then head outside. Veteran astronauts who have stepped out into space this way suggest that first-timers pause for a half minute to burn the image of what they see into their memories. From inside the shuttle, the view through the multi-paned windows isn't 100 percent clear. In a spacewalk spacesuit, known as an extravehicular mobility unit or EMU, there's just a thin clear visor separating the wearer from the deadly vacuum of space. The view is spectacular, so spacewalkers are encouraged to enjoy it.

The hazards of venturing outside the shuttle were a major issue during the deliberations on how to solve the Intelsat rescue. "We sent a couple of ideas to the ground," says Brandenstein. "I don't think the three-man spacewalk was really our top option. We had another plan that called for just two spacewalkers, but we finally decided that the three-man plan was the way to go."

Convincing Mission Control meant covering issues of survival. If, for instance, a spacewalker runs an oxygen loss, he or she needs to be able to "buddy breathe" off a partner. This requires elbow room to pass the air umbilical back and forth—which makes three people in spacesuits in the airlock a possible risk.

That was just one concern in *Endeavour's* plan to send three people out. The scenario called for the three spacewalkers to perch on the sill of the cargo bay while the shuttle edged close to Intelsat. The astronauts would reach up with their gloved hands and grab the satellite. The danger there lay in the sharp-edged titanium on the bottom of Intelsat, which could cut or puncture a spacesuit. Brandenstein felt that was one of the lesser hazards in the three-person spacewalk plan, given the crew's prelaunch training on the ground. Astronaut Pierre Thuot had done work with a pair of spacewalk suit gloves inside a device called a glove box, which could create

Figure 23. Artist's conception of the Galileo space probe at work around the planet Jupiter. This rendering doesn't include the stuck main antenna, which was supposed to open fully like an umbrella after deployment from the shuttle. It unfurled only partway, limiting the speed at which data and photos of the giant planet were transmitted to Earth. Courtesy of NASA.

the vacuum he would encounter outside the shuttle. "Pepe [Thuot] had a piece of titanium like on the bottom of Intelsat," recalls Brandenstein. "And he started hacking away at the glove with it and never did damage it." The big concern was fitting all three spacewalkers into the airlock and getting them in and out safely.

Four EVA-qualified astronauts were aboard *Endeavour*, Thomas Akers and Kathryn Thornton as well as Thuot and Hieb. Thuot, Hieb, and Akers were selected to make the risky EVA because they were of similar height and wore similar suits. Akers and Thornton would later distinguish themselves as spacewalkers on the shuttle mission to repair the fuzzy vision of the Hubble Space Telescope.

Dan Brandenstein was worried that Houston was going to think its way out of the three-man scenario, which the *Endeavour* crew felt was the best way to go. "They [Mission Control] were going to get three people and put them in the water tank," says Brandenstein. "I was afraid they were going to come up with the wrong answer." The tank that *Endeavour's* commander was referring to was the Weightless Environment Training Facility or WETF. Spacewalkers put on their suits and plunge into this pool of water to simulate the weightlessness of space. The *Endeavour* crew wanted to head off what they felt might be a bureaucratic mistake. "We got three people we wanted to send out, and we suited them up and put them in the airlock to make sure they could pass the umbilical back and forth. Obviously, they fit." Houston said okay, and Akers, Thuot, and Hieb were given the go-ahead to venture out in the payload bay. Brandenstein's first message to Mission Control said it all.

"Houston, I think we've got a satellite."

They did.

Years after Intelsat 6 had been tucked into its correct orbit, astronaut Mike McCulley was flying to Houston on a commercial passenger jet. It was 1996, ten years after the *Challenger* accident. The woman sitting next to McCulley asked what he did for a living. The veteran astronaut of the STS-34 mission aboard *Atlantis* couldn't help but notice his neighbor was reading a story in *USA Today* with photographs of the planet Jupiter taken by the Galileo spacecraft that he and his astronaut crewmates had deployed in 1989. "That's what I do," said McCulley, pointing to the paper. It seemed like NASA had bounced back from the loss of *Challenger*. The agency still didn't have a mission, or a space station of its own, but it could borrow both from someone else.

1993

The Road to Mir

The image of three spacewalkers from space shuttle *Endeavour* grabbing and saving the Intelsat 6 satellite gave NASA a much needed jolt of good publicity. Astronauts were seen doing something that unmanned rockets couldn't do. For the beleaguered U.S. space program, it was the highlight of 1992. The shuttle still didn't have a specific mission to perform, but it was about to get one. However, international diplomacy would be the driving motivation, and not necessarily science or exploration.

When William Jefferson Clinton was inaugurated in 1993 as the forty-second president of the United States, the new administration inherited more than just the keys to the Oval Office. There were also questions on what to do with NASA, and a new White House science advisor needed to be hired to talk out those issues with the new commander-in-chief. The result would be what seemed like the right idea pursued for the wrong reason.

After working with the Nixon administration in 1973 as the first chief of the Department of Energy Conservation and years with the Office of Technology Assessment, Dr. John Gibbons found himself sitting down with the president-elect. "We talked for about an hour and a half," says Gibbons. "We didn't talk about the job, just on my science philosophy and his." Clinton gave his new science advisor one basic guideline. The president admitted that he and Vice President Al Gore were both wonks and that Gibbons shouldn't be worried about giving them too much to read.

Mr. Clinton felt that science might be the way to solve social problems, but there was also the deficit and the end of the Cold War to deal with. The demise of the Soviet Union meant the United States didn't need the kind of military expenditures President Ronald Reagan once advocated to fight what he termed the Evil Empire. That foe was gone, and states like California, Texas, and Florida were in jeopardy of losing aerospace jobs. Those states also accounted for a hefty portion of the electoral votes Mr. Clinton would need for a second term.[1]

According to John Gibbons, President Clinton wasn't an unabashed fan of manned spaceflight. "We didn't talk about it much," he says. However, the newly elected chief executive did see an opportunity to kill two political birds with one stone. Mr. Clinton could advance diplomacy with the Russians and deal with a pair of science projects known within the walls of the White House as the "two enchiladas." One was an atom smasher called the Super Conducting Super Collider, and the other was space station Freedom. Both were leftovers from the Reagan and Bush administrations, and now they were in President Clinton's lap.

One of John Gibbons's first major ventures on Capitol Hill was to lobby lawmakers for money for the atom smasher. Gibbons and Clinton met beforehand for two hours in the Roosevelt Room to discuss the situation. Half of the meeting was dedicated to the collider and the second hour to space station Freedom, which would be built and maintained by the U.S. space shuttle. The two talked about international participation in the atom smasher to lower the cost and make the project more palatable to Congress. They decided there wasn't time to negotiate a deal, so Gibbons went before the House to suggest the United States go it alone. "I went to Congress and made an impassioned plea on the expected and unexpected benefits of the project," he says.

Congress proceeded to kill the first "enchilada."[2]

That lesson sank in quickly within the Oval Office and had a direct impact on what to do about the space station. The need for foreign participation became apparent, and that meant inviting Moscow to the table. Europe, Japan, and Canada had signed on to participate, but more help was needed to sell Congress. "If we could internationalize it, we knew the Russians had a lot of experience," says Gibbons. "It could have multiple benefits. It was one of the things the Russians still prided themselves on, and they could bring something to the table." Just as NASA had felt it would have a better chance to create its troubled space shuttle program if it had the military signed on as a participant, the Clinton White House felt it would fare better with the space station if it was standing shoulder to shoulder with Russia. However, Gibbons also acknowledged that persuading Congress to allow former Cold War foes onto the hallowed ground of the Kennedy Space Center and possibly onto missions of the shuttle would be difficult. Concern was also being voiced that farming aerospace work out to Russia in a possible joint space station could cost American jobs.

NASA, for its part, knew who the real expert was when it came to making space stations that worked during long-term stays in low earth orbit. Shuttle crews came back from mere weeklong missions like triumphant heroes. By the time of the *Challenger* accident, the Soviet Union had already launched seven of its Salyut stations and the core of the space station Mir (meaning "peace"), where cosmonauts would spend up to a year in orbit.

Reviving an Old Space Partnership

In 1975 a thaw in the Cold War between the United States and the USSR led to an agreement between President Richard Nixon and Soviet leader Leonid Brezhnev to stage a joint mission between the two superpowers' space programs. Dr. Roald Sagdeev saw these events firsthand. From 1973 to 1988 he was director of the Space Science Institute in Moscow. He would eventually serve as a space advisor to the Soviet Union's last president, Mikhail Gorbachev. He would later join the Eisenhower Institute in Washington, D.C. "Brezhnev and Nixon were in a hurry to find some kind of spectacular demonstration in orbit for the era of detente," says Sagdeev. "Substance was almost absent."

The scientific community in the Soviet Union was hoping for something more ambitious, like a joint expedition to Mars. Going back to the moon seemed less attractive because of the resounding defeat Moscow had suffered at the hands of Washington when Neil Armstrong and Buzz Aldrin landed at the Sea of Tranquility on the lunar surface. A well-worn Russian saying is sometimes used to explain why. Loosely translated, it goes: "In the house of the hanged man, you don't talk about rope."[3]

"The scientists [in Moscow] were told, don't worry, you will have your chance," says Sagdeev.

That chance meant more scientific capability once it was proven that Apollo capsules from NASA and Soyuz vehicles from Moscow could dock and operate jointly. Sagdeev and his colleagues were satisfied for now.

NASA selected three men to fly the Apollo side of the mission. Astronaut Tom Stafford had flown to within a whisker of the moon from *Apollo 10* and then pulled away in the lunar module Snoopy as the dress rehearsal for Armstrong and Aldrin on *Apollo 11*. Stafford was joined by Donald "Deke" Slayton, who had regained his flight status after a heart murmur left him on the sidelines for much of the Apollo program. The third member was Vance Brand, one of the astronaut candidates left with no chance to walk on the moon when NASA's budget was cut. He was supposed to fly to the moon as command module pilot on the cancelled *Apollo 18* mission.[4] When the moon program was scrapped, Brand stuck it out at NASA until the shuttle started flying in 1981. In 1982 he was made commander of STS-5, the first shuttle mission to carry commercial satellites and crew members trained to perform spacewalks.

In 1975 the Apollo-Soyuz Test Project featured the first docking of a U.S.-built Apollo spacecraft and a two-man Soyuz craft from the Soviet Union. It took twenty years for a follow-up partnership to occur. Apollo-Soyuz helped paved the way for the space shuttle to dock with the Russian space station Mir.

Astronauts Brand, Stafford, and Slayton would launch aboard the last Apollo, propelled by the last Saturn rocket. In the Soyuz capsule, rookie Valery Kubasov teamed with a Soviet space legend. Cosmonaut Alexei Leonov, who had made the world's first spacewalk, commanded the Soviet half of the joint mission. The Apollo crew knew they were working with a celebrity with a jovial personality and a good sense of humor.

That's not to say that NASA, the astronauts, and the press immediately

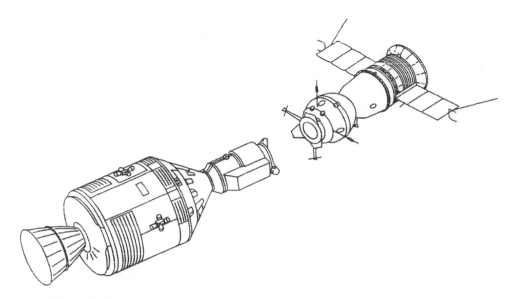

Figure 24. The Apollo-Soyuz Test Project in 1975, featuring the first-ever docking between U.S. and Soviet spacecraft. NASA's last Apollo capsule carried a docking module to orbit so it could link up with the Soyuz. Astronauts Deke Slayton, Vance Brand, and Tom Stafford exchanged greetings with cosmonauts Alexei Leonov and Valery Kubasov. No other joint U.S.-Russian dockings would occur until 1995. Courtesy of NASA.

warmed up to a joint flight with the Soviets. Working with communists was only one issue; the Soyuz craft itself was another. The workhorse of the Russian space program was considered a marginal vehicle at best. The *Soyuz 11* disaster was just one example of how dangerous Soviet space travel could be. About the time of NASA's *Apollo 15* lunar landing, the Soviets sent cosmonauts Georgi Dobrovolski, Viktor Patseyev, and Vladislav Volkov on the first mission to the Salyut 1 space station. Following touchdown, the crew remained silent inside their capsule. Recovery teams soon found out why. During the trip down, an air valve on the Soyuz came open and the atmosphere inside the crew cabin rushed out. The cosmonauts weren't wearing spacesuits because there were three of them and the Soyuz was cramped. As a result, they died of asphyxiation.[5]

The primary crew for this ill-fated mission was supposed to include Alexei Leonov and Valery Kubasov, but the backup crew went instead, allowing Leonov and Kubasov to be assigned later to Apollo-Soyuz. While

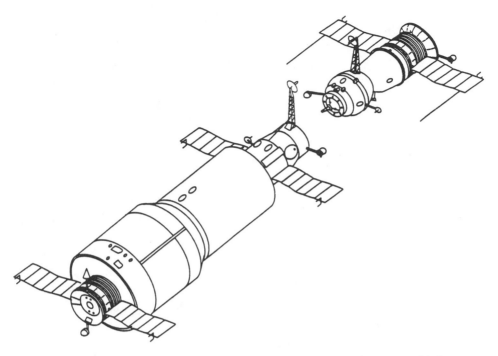

Figure 25. Salyut 1, the first of the Soviet Union's space stations that operated in low earth orbit. Soyuz space capsules would dock to the forward docking port. There was no rear port, so resupply ships couldn't bring up food, water, and fuel. Later Salyuts had a rear docking port for cargo deliveries. Courtesy of NASA.

the Americans regarded the Soyuz with trepidation, Moscow's view of NASA's Apollo spacecraft was awe tinged with a little intimidation. The Saturn V rockets that took astronauts to the moon were considered big by the Soviets. So was the Apollo command module. "I would say the overwhelming impression was of how huge it was," says Roald Sagdeev. "The Soyuz was really tiny compared to Apollo. Even our later Salyut space station was smaller."

Docking a 60,000-pound Apollo craft to a 6,000-pound Soyuz presented obvious technical challenges. Further, the air pressure inside the Apollo command module was lower than in the Soyuz. The solution was the docking module that was carried to orbit by the Apollo spacecraft as the third piece of the puzzle to link the two space vehicles. The two-ton, ten-foot-long compartment would connect the Apollo and Soyuz vehicles and equalize the air pressure to allow visits among the crew members.

The docking mechanism did away with the probe-and-target design used on Apollo. It also dispensed with the more political question of who was "active" and who was "passive" during this mission among equals.[6]

That ended the big technical questions. The cultural and political ones remained. The Soviets had heard a lot of things about Americans, and the astronauts had heard a lot about the Russians. Very little of it was complimentary.

The Apollo astronauts made trips to Moscow, posed for pictures, visited historic places, and learned the Russian way of doing things in space. Cosmonauts Leonov and Kubasov returned the favor by flying to Texas to watch calf roping and eat barbecue. The astronauts learned basic Russian, and the cosmonauts learned basic English. During all the preparation for the joint docking mission, Roald Sagdeev saw some of the fakery in the name of national security.[7] The head of the Space Science Institute says he was asked to act like he ran the whole show on the Soviet side. The reason was that much of the Russian space program at that time was operated by the military, and the notion of having Cold War adversaries poking around didn't sit well with many of the generals that Sagdeev dealt with. Part of the strategy was for the military to keep a low profile and encourage Sagdeev to be the face the Americans saw. That included a quick change of clothing for the generals. "When the Americans appeared, these guys would put on civilian clothing," says Sagdeev. "One general asked, 'When will your friends [from NASA] be leaving? These clothes are uncomfortable.'" Everyone in the Russian space program pretended it was a civilian agency from stem to stern and operating through the Space Science Institute.[8]

The countdown eventually proceeded, and the Soyuz rocket, with Leonov and Kubasov on board, blasted off seven hours before the Saturn carrying Brand, Stafford, and Slayton. Months of computer simulations didn't make the final approach and docking any less strange for the astronauts. The Soyuz appeared bulbous and buglike and covered in green fabric. Docking took place, and Slayton and Stafford sealed themselves inside the docking module so the air pressure could be raised. Stafford opened the hatch from the American side and Leonov did likewise from the Soviet side. Joint documents were signed, souvenirs were exchanged, and the mission was declared over. Reentry into the earth's atmosphere, however, didn't go as planned. Radio interference kept Brand from hearing the command to switch off the Apollo capsule's automatic thrusters,

Figure 26. Artist's conception of a docking between the space shuttle and the Salyut space station. This mission, intended as a follow-up to the Apollo-Soyuz Test Project, never took place. The shuttle later docked with the Russian space station Mir. Courtesy of NASA.

which used two toxic chemicals. Some of the nitrogen tetroxide leaked into the cabin through a vent, leaving Brand unconscious and the crew wearing oxygen masks. The corrosive chemicals left blisters on the inside of the astronauts' lungs.[9]

This close call notwithstanding, talks continued on future joint missions between the United States and the Soviet Union. One plan would involve the NASA space shuttle that was still in development. The Soviets suggested docking the shuttle to one of its Salyut space stations. If that worked, scientists like Roald Sagdeev at the Space Science Institute wanted to build an experiment module similar to compartments that would eventually be launched for the International Space Station. "At that time, we were disappointed with the scientific value of Apollo-Soyuz," says Sagdeev. "So scientists from both communities were asked to come up with interesting ideas."[10]

Then things started to change.

Watergate had forced Richard Nixon from power, leading to the administrations of Gerald Ford and Jimmy Carter. Detente cooled between the Americans and the Soviets, and NASA and Moscow slowly went their separate ways.

The Soviet Union Builds Its Specialty of Space Stations in Low Earth Orbit

The first Soviet stations were called Salyuts, as a "salute" to cosmonaut Yuri Gagarin.[11] Each would perform a series of missions and then be abandoned to descend unmanned into the atmosphere and burn up. Moscow was already using Salyut 4 by the time Apollo-Soyuz blasted off.

The Soviets were learning practical lessons on long missions in orbit. One bit of experience concerned resupplying the outpost. Salyuts 1, 2, 3, 4, and 5 had only one docking hatch. The Soyuz vehicle carrying the cosmonaut crew and their supplies would dock to the Salyut. The problem for arranging longer flights was becoming apparent. Extra provisions couldn't be sent up, because there was no place for the arriving craft to dock. Also, the design of the Soyuz spacecraft meant it could last only about six months in the cold of space. After that, the ship's fuel lines would freeze up and the vehicle couldn't safely return the cosmonauts to Earth.

Moscow's solution was easy. They added a back door.

Salyuts 6 and 7 had a rear docking port, which opened a number of opportunities for the Soviet space program. Number one was the ability to receive supplies on automated cargo ships called Progress vehicles, which were basically Soyuz craft with the pressurized crew compartment removed and replaced with warehouse space for food, water, and gift packages from the cosmonauts' families. The Progress ships were designed in such a way that when the craft docked to the Salyut, fluid lines leading

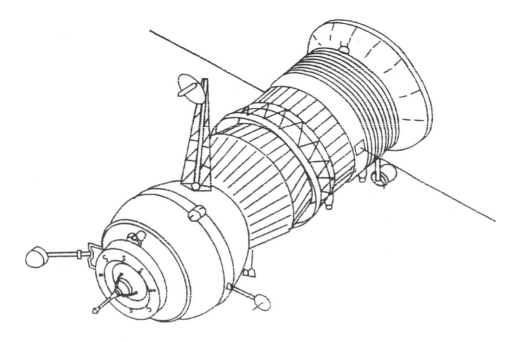

Figure 27. The Russian-built Progress vehicle, an unmanned version of the manned Soyuz craft. The pressurized crew module is replaced with a small cargo compartment for food and gift packages from the cosmonauts' families. Water and fuel tanks can pump their contents into the station after docking. Progress 1 flew to the Salyut 6 station in 1977. Courtesy of NASA.

into the station would automatically connect so the liquid cargo could be pumped in.[12]

The second docking port also solved the problem of the Soyuz's longevity. If a Salyut crew was assigned to stay in space longer than the six months its Soyuz could last, another Soyuz would be sent up on a "taxi flight." A trained cosmonaut pilot would dock the new Soyuz and, after a short visit, return to Earth in the older Soyuz, leaving the Salyut crew with a fresh vehicle. The six-month clock then started all over again.

A unique political opportunity accompanied this solution. The visiting pilot could be accompanied by a second crew member, who might be a guest cosmonaut from a Warsaw Pact nation. The door was also opened to so-called Intercosmos flights, where guest researchers from Cuba, Vietnam, and even Mongolia could go into space on missions that helped cement ties between the Soviets and their allies.[13]

The possibility of docking a U.S. space shuttle to a Salyut station was also rekindled. By the time of the 1985 summit between Mikhail Gorbachev and Ronald Reagan, Moscow wanted to talk about even more ambitious joint space missions than Apollo-Soyuz. Again Roald Sagdeev sat in on the meetings, and found himself in the middle of the discussion of a joint Mars mission during a state dinner in Moscow in honor of President Reagan. Sagdeev was in the receiving line to greet the leaders when Gorbachev spotted him. "He grabbed my hand," says Sagdeev, "and he said to Reagan, 'This is the man who is dreaming of a joint mission to Mars,' and then he kissed my hand." Reagan appeared to have a twinkle in his eye following the exchange, but no Mars trip resulted. More progress wouldn't occur until after the breakup of the Soviet Union in 1991.

After that, democratically elected Russian president Boris Yeltsin appeared in the East Room of the White House in 1992 to sign seven documents with President George H. W. Bush. One dealt with the elimination of some nuclear weapons; another was about space cooperation. All of the lengthy negotiations leading up to the pact on U.S.-Russian space cooperation started people talking at the Johnson Space Center in Houston.[14]

Astronaut Norman Thagard didn't give much thought to the rumors that an American would visit the Mir station. He figured it was being handled by political heavy hitters from both nations. In late 1991 he had a shuttle mission to train for, teaming him with Ron Grabe, Steve Oswald, Bill Readdy, Ulf Merbold of Germany, Roberta Bondar of Canada, and Thagard's old officemate at the Johnson Space Center, Dave Hilmers. This was the International Microgravity Laboratory inside a Spacelab module that would fill the entire cargo bay of space shuttle *Discovery*.

It was when Thagard was training for his mission at the Marshall Space Flight Center in Alabama in late 1991 that he missed a crucial meeting called by the chief of the Astronaut Office in Houston. "Dan Brandenstein met with everybody and said a joint mission was coming on Mir," says Thagard. "He asked if anybody was interested in going." The talks were in their early stages between Moscow and the first Bush administration, so it would be some time before an astronaut was picked for the flight, which would pair one U.S. crew member with two Russian cosmonauts for several months aboard the orbiting Mir station. That didn't keep more than a dozen shuttle crew members from volunteering for the job. Thagard wasn't among them, since he hadn't heard that the vague rumors were about to become someone's ticket to soar into orbit inside a Soyuz capsule. That was something even the Apollo-Soyuz crew couldn't claim.

Months passed before the subject came up again. The list of volunteers sat gathering dust, and Thagard was considering his future while sitting in the Houston office he shared with Dave Hilmers. Each man had been in space four times on notable missions, and making a fifth began to sound redundant. Thagard's first flight had been with Sally Ride, America's first woman in orbit. He followed up that 1983 trip aboard *Challenger* with a Spacelab science mission on the same shuttle in 1985. In 1990 he was part of the crew that launched the Magellan probe to Venus, the first interplanetary craft carried up by an orbiter. Then it was eight days in space with Hilmers in 1992.

His officemate's résumé was equally cluttered with experience in orbit. Hilmers was on the maiden flight of *Atlantis* in 1985, plus a classified military mission and the trip on *Discovery* he shared with Thagard. Perhaps his most notable blastoff was aboard the first flight to go after the 1986 *Challenger* disaster.

The two men began talking about what was next for them, and space wasn't necessarily it. Hilmers was then considering going to Baylor University to study medicine, and it would take something more than another space voyage to derail those plans. Thagard had been thinking about the future himself, and the destination wasn't orbit. It was Tallahassee. Specifically, there was a possible offer to teach engineering at his alma mater, Florida State University.

"We'd been talking about it on and off," says Thagard. "And I told Dave the only thing that might keep me from taking it was going to Mir."

The matter might have died right there, had one of Thagard's former crewmates not been standing in the door. Astronaut Mark Lee was also on the shuttle mission carrying the Magellan probe, and he had news to offer. Word that NASA was really looking for someone to go to Mir sent Thagard to the chief of the Astronaut Office, Dan Brandenstein. But when Thagard saw the list of volunteers, he thought that was that.

Weeks passed before the subject came up again. That time, Brandenstein was more direct and called Thagard into his office to offer him the mission. Thagard was of two minds. He remembered the days of the Apollo-Soyuz Test Project, which occurred during a temporary pause in the Cold War. But nothing happened following that mission for the next twenty years. The gap puzzled Thagard. Then one day, when he was going through his slide collection, he came across some shots of himself standing next to the F-4 Phantom jet fighter that he flew as a captain in the U.S. Marine Corps during the Vietnam War. A few years after the conflict, a Vietnamese cosmonaut would visit the Salyut 6 space station during the Soviet Intercosmos flights. Now Thagard might be a guest on a similar mission. "The irony of that is that I would be under the command of a Russian fighter pilot," says Thagard. "And that's not how I thought that kind of encounter would be."[15]

The political ball began rolling when President Bush signed the deal for joint space projects in 1992, but it was President Bill Clinton who implemented this idea when he moved into the White House a year later. Thagard was soon training to squeeze into a Soyuz space capsule for the trip to Mir. It would be the first of a series of flights building up to the International Space Station. Thagard began training for his mission to Mir, not in Moscow, but with his nose buried in a first-year college textbook on basic Russian language. After that, he attended classes at the Defense Department Language School in Monterey, California. Then he was ready to train to fly on a Soyuz rocket. NASA didn't have money in the budget for all the language lessons, so Thagard pulled out his checkbook and picked up the difference. When he went, he wanted to be ready.

Macy's Meets Gimbels, NASA Style

NASA would be working with the Russians for the first time since Apollo-Soyuz in 1975, and not everyone was comfortable about it. By docking the shuttle to Mir as a test bed for building the International Space Station, NASA appeared to be giving the shuttle the true mission it had lacked for years. Still, critics in the United States complained that space policy was becoming foreign policy and that joint missions with the former superpower were little more than foreign aid with spacesuits.

Concerns over the joint space station project were also being voiced in the former Soviet Union, but they were a bit different. Support for the Russian space program was fading as well. The new joint space station was seen as a way to preserve both aerospace jobs and aerospace technology following the fall of communism.[16]

Most Russians could be counted on to know how long the Mir station had been in orbit, but enthusiasm among Russian youth wasn't what it once was. "If you ask young men, they would not remember Yuri Gagarin," observes Roald Sagdeev. "John Lennon and Paul McCartney maybe. The Beatles left a bigger impact on the minds of young people than Sputnik." Plans were made for joint missions, and space shuttle flights were scheduled to take the baby steps leading to the construction of the International Space Station.

NASA had a general idea of how the new outpost would be built in orbit. The cluster of crew modules and other pieces had to be prefabricated, built in chunks that could fit in the cargo bay of the space shuttle. The design had a long spinelike framework where the crew cabins would be attached. The new station would need electricity, and that meant long winglike solar power panels to convert sunlight into energy. Those parts would have to be bolted to the station's frame as well.

Spacewalkers in 1985 had tried to float outside the shuttle and assemble a practice frame by hand, which didn't work. The actual structure would need power and computer data cables and coolant lines running through it like spaghetti, and that would be too complicated for people to do while in orbit. All of these parts would have to be fully built on the ground and taken up on successive shuttle flights.

None of this would be possible if the shuttle couldn't rendezvous and dock with the orbiting complex, and that's where the joint flights aboard the Mir station proved invaluable. Shuttle crews could practice many of the skills they needed by docking with Mir. Then NASA could build the new outpost. Baby step number one was a rendezvous in early 1995 by space shuttle *Discovery*.

Veteran astronaut Jim Wetherbee led the six-member crew, which included the first woman to act as copilot of the shuttle, Eileen Collins. Michael Foale, Janice Voss, and Bernard Harris, also veteran astronauts, rounded out the crew along with cosmonaut Vladimir Titov. Titov's job was to operate the radio during the close approach and to speak with his colleagues aboard Mir in rapid-fire Russian.

After a trouble-free launch, a leaking aft jet thruster on *Discovery* almost prevented the crew from getting close to the station. Skittish ground controllers in Moscow were worried over the damage the shuttle might do if the astronauts sprayed corrosive jet fuel on the solar electricity panels that stuck out from the space station. The two sides conferred, and a deal

Figure 28. Astronauts Jim Wetherbee and Eileen Collins talking following the landing of space shuttle *Discovery* after the first rendezvous between the U.S. shuttle and the Russian space station Mir. Collins became NASA's first female shuttle pilot and went on to become the first female commander in 1999. Courtesy of NASA.

was struck to close off the fuel supply to the leaky thruster and allow the shuttle to come within thirty feet.[17]

The maneuver began on the dark side of the earth, with the planet cutting off light from the sun. All the astronauts could see was a single floodlight blinking rhythmically in the distance. The eerie scene continued as the two vehicles drew closer together, and the outpost slowly came into focus. A white central crew compartment with the blinking light was connected to a short, stubby, boxlike module at the back, a long module sticking up, and another sticking down. Each long module had a pair of solar power panels sticking out to the left and right, forming neat rows. Docked to front and back were two of the Soyuz-type vehicles that the Apollo-Soyuz astronauts had described as bulbous, buglike, and covered with green fabric.[18] At the center of the main crew compartment was a single round window. Peering out was the face of cosmonaut Elena Kondakova. "How can we be doing this? This is incredible!" said *Discovery* commander Jim Wetherbee after the shuttle's landing. He sounded like a man who just had seen a flying saucer land in his backyard and wanted the whole world to know about it.

The rendezvous between space shuttle *Discovery* and Mir cleared the way for Norman Thagard's long-duration stay on the complex. *Discovery*'s flight also set the stage for the next shuttle visit to Mir. This follow-up mission would involve docking the two spacecraft together, and the commander of this planned *Atlantis* flight, astronaut Robert "Hoot" Gibson, watched *Discovery*'s progress closely.

Like Thagard, Gibson was one of NASA's original shuttle crew members hired in 1978. He was pilot on the first shuttle to land back at the Kennedy Space Center instead of the dry lake bed of California's Mojave Desert. Later he would be commander aboard the last shuttle to fly before the *Challenger* disaster in 1986. Gibson says his third mission, which was

Figure 29. The STS-71 crew heading to the launchpad for the first docking flight between the space shuttle and the Russian space station Mir. Commander Robert "Hoot" Gibson leads the way, followed by astronaut Charlie Precourt and, at left, cosmonauts Nikolai Budarin and Anatoly Solovyev. At the back, left to right, are astronauts Ellen Baker, Greg Harbaugh, and Bonnie Dunbar. Courtesy of NASA.

a classified flight for the Pentagon, gave him critical training for his flight to Mir. The now declassified mission included a rendezvous. "I think every shuttle pilot dreams of a rendezvous," says Gibson. "There were a lot of fellows who would have been able to do this [Mir] flight, and NASA decided it would be me." To be sure, catching up with a Defense Department payload the way Gibson did on his earlier flight might not be as challenging as linking up with a hundred-ton space station.

The plan called for the astronaut to float to the rear of *Atlantis's* flight deck and use two small joysticks to fire the shuttle's thrusters during final approach. His view of the whole operation would be through the four windows overlooking the payload bay and out toward the target station. *Atlantis* would have to glide toward Mir at about one inch per second and then dock within a two-minute timeframe so the two spacecraft would be high over a Russian ground station for good radio contact. Compared to his other space missions, Gibson says, nothing came close to this. "I would call this the toughest, the most exciting and ambitious. And I've done some ambitious things on my earlier flights." One of the primary tasks of Gibson's flight was to bring astronaut Norm Thagard back to Earth. But the first American to live and work aboard the Mir station had to get there first, by taking a seat on a Russian Soyuz rocket.[19]

The First American to Launch on a Rocket since 1975

The successful rendezvous between the shuttle and Mir gave NASA the confidence it needed to okay Thagard's blastoff one month later to visit the outpost. Still, the official agreement for his mission guaranteed the astronaut a flight back to Earth on a Russian Soyuz if the shuttle was unable to dock with Mir for his return trip.

Thagard would be the first American to launch on a vehicle other than the space shuttle since 1975. The Russian Soyuz had evolved since its first disastrous flight in 1967, in which the one cosmonaut on board was killed in a crash landing. The newest model of the capsule was called Soyuz TM, for "transport modified," and that's the one Thagard would fly. The cramped middle compartment was the crew cabin where the astronaut would sit with his knees almost to his chest during the trip from Earth to Mir. The shuttles Thagard flew aboard during his four space missions still had the clunky mechanical gauges in the cockpit known to pilots as "steam

Figure 30. NASA astronaut Norman Thagard sitting with Russian cosmonauts Vladimir Dezhurov and Gennady Strekalov. The three men flew to the Mir space station aboard Soyuz TM-21 and would later return to Earth aboard space shuttle *Atlantis.* Courtesy of NASA.

gauges." The dashboard of the Soyuz would be somewhat similar, with square buttons colored in white, blue, and red. There were also knobs and valves, all within inches of the crewmen strapped in their custom-made flight couches. Of course, the controls were marked in Russian. "I thought it was great, just like any vehicle you're going to go up in," says Thagard. "It's quite an experience to get to rummage around in it."[20] The front compartment, which gives the Soyuz its bulbous appearance, is the docking module, which is equipped with a docking collar to attach the Soyuz to the Mir station. The rear third of the craft is an instrumentation compartment containing the ship's twin solar panels.

Getting ready for the mission meant Thagard had to become proficient on the Soyuz, even though the piloting of the cramped capsule would done by cosmonaut commander Vladimir Dezhurov and veteran flight engineer Gennady Strekalov. Thagard's training included squeezing into a Russian-made Sokol pressure suit. The one-piece visor can slide into place and lock only when the cosmonaut is seated, uncomfortably, in the custom-fitted flight couch with knees drawn up tightly. Critics often say

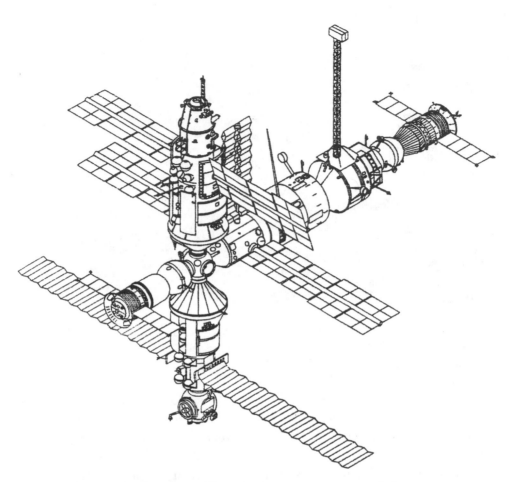

Figure 31. Configuration of Russia's Mir space station around the time astronaut Norman Thagard arrived in 1995. The station pictured here features the main Mir crew compartment, the Kvant 1 and Kvant 2 modules, and the Krystal compartment where the space shuttle would later dock. The Spektr and Priroda modules are absent. Courtesy of NASA.

NASA's space shuttle is as roomy as a pop-up camper. But it seemed cavernous compared to the Soyuz where Thagard would spend the two days between launch and docking.

The crew module that the Soyuz crew would occupy for the fiery trip to orbit was the middle section of the three-part space capsule. Thagard had made the trip to space four times before on the solid fuel boosters of the shuttle. After his Soyuz had been delivered, in one piece, by railcar to its launchpad in Kazakhstan, the crew would squeeze inside and wait out the countdown.[21]

The ride to orbit, for Thagard, wasn't all that different. "The noise and vibration and the popcorn popping [sound from the shuttle's solid rocket boosters] was louder. But the force of gravity wasn't all that different. Although, the separation of the third stage of the Soyuz was pretty emphatic." The cramped space capsule would coast around the globe for two days before docking with the Russian space station Mir. Thagard's best view of the station wouldn't come until months later when space shuttle *Atlantis* docked with the outpost and he could look out the shuttle's windows. The initial trip toward Mir aboard the Soyuz afforded almost no view at all. "I had a window," Thagard says, "but it looked out to the right away from Mir. The commander had a small periscope to look through, but all I could see was a pale blurred image." The astronaut would trade the view of his Soyuz crew cabin for the inside of the capsule's docking compartment, then a hatch would open into the modules of Mir.

The Mir crew welcomed Thagard with the traditional gifts that visitors receive in Russian homes. The list included chocolate, bread, and salt. The welcoming committee was made up of the three people who had been aboard Mir when *Discovery* made its historic rendezvous, cosmonauts Alexander Viktorenko, Valery Polyakov, and Elena Kondakova. Thagard got his first impressions of the Mir station that America had heard about for a decade. He would be the first member of the Soyuz crew to have a look around. After entering the station through the smaller Kvant compartment, where the Soyuz docked, Thagard floated into the main Mir crew cabin. The astronaut wasn't seeing the station at its best, since these were the oldest of the modules. "It looked like the inside of someone's utility room," he says. "You could tell it had been there for a few years, and stuff had been stuffed here and there and stored away."

The central Mir crew compartment was an improved model of the Salyut 6 and 7 stations that flew in orbit in the late 1970s to mid 1980s. The Kvant 1 compartment (the word means "quantum") would be launched next and dock to the rear of the Mir compartment. After that came the boxcar-sized Kvant 2, which included an airlock for spacewalks and a shower for the crew. Last was the Krystal module, which would play a crucial role in future joint missions with the space shuttle. It had a docking collar originally designed to work with a Soviet version of the U.S. shuttle, which was called Buran ("snowstorm"). Space shuttle *Atlantis* would be outfitted with a similar collar to link itself with the outpost.[22] There were six people aboard Mir for only about a week. Then Viktorenko, Konda-kova, and Polyakov boarded their Soyuz TM capsule docked to the front port on the station and headed back to Earth. After that, Mir would seem even roomier for Thagard and his two crewmates.

Thagard's four months on the orbiting space station would break the previous U.S. endurance record set by the third Skylab crew in 1974. Still, the visit wasn't without its snags. One involved food. Mission planners in Moscow wanted to carefully monitor what the crew ate and drank during their time in space. Accordingly, a galley on Mir was stocked with pack-aged food items marked with bar codes. Before eating, the crew would scan the cans to keep a running tally. At least, that was the plan. But no one asked the crew what they wanted on the menu. Thagard was used to the cuisine on the space shuttle, which was mostly tailored to the tastes of each crew member. Worse, on Mir four days out of every six featured canned fish. "I don't eat canned fish, I just don't," says Thagard. "Fish has to be prepared very nicely for me to eat it at all, so the upshot was that there were things in those six-day meal plans I wouldn't eat." Not every dinner is a big hit aboard the shuttle, but at least astronauts can resort to the "pan-try." That's a locker containing things like peanuts that crew members can snack on between meals. The Mir station had a pantry with supplemental food to choose from, but these items didn't have the bar codes Russian mission controllers wanted the crew to scan. Thagard had the option to write down all the junk food he chose to eat, except that there was almost no scrap paper on Mir to make notes. "I stuck with the bar-coded food and lost seventeen pounds," he says.

The astronaut's job was as a researcher with a list of experiments to conduct. But he was also taking notes on how Mir worked, as NASA went about designing its portions of the planned International Space Station. The Russian outpost had been in orbit for nine years, but despite such longevity, Thagard immediately began to notice things that shouldn't show up on a new space station. For instance, long coiled air vents snaked from module to module, which could make it tougher to close off a hatch if one of the Mir modules sprang an air leak.

Also, storage was a concern. There was almost no place to tuck away tools or supplies, so keeping track of things on the station could be a bit slipshod. Often the only person to know the location of a certain tool was the last person to use it. Things got even more complicated if that cosmonaut had already returned to Earth. Moscow's answer was to make a phone call. "When I couldn't find some components for a medical test," says Thagard, "they called Valery Polyakov at home—he lived in Moscow at the time—to ask if he knew where this particular item was."

The astronaut's plan to conduct life-sciences experiments was cut back when the launch of Mir's latest module, called Spektr ("spectrum"), was delayed. This science compartment had two sets of solar panels to add to Mir's power supply. Its February launch was deferred until June, leaving Thagard only about a month to do experiments before space shuttle *Atlantis* docked to bring him back to Earth. In the meantime he quickly ran out of things to do. "It was boring," says Thagard. "You can actually get bored in space." He tried to fill his time doing earth observation and taking photographs of the planet as Mir sailed around the globe. There was another way for Thagard to pass the time, but his conscience kept him from indulging in it. "I had a *New York Times* crossword puzzle book," says Thagard. "But while I didn't have much to do, my crewmates had too much work. I just couldn't bring myself to work crossword puzzles while they went about busily working. I just kept taking earth observation pictures." Then June brought Spektr, and all too soon Thagard's flight aboard Mir was due to end with the upcoming docking of the space shuttle *Atlantis*.

The shuttle crew that would accompany "Hoot" Gibson on this mission in July of 1995 included copilot Charlie Precourt, astronauts Greg Harbaugh, Ellen Baker, and Bonnie Dunbar, and cosmonauts Anatoly Solovyev and Nikolai Budarin. All of the *Atlantis* crew would build on the experience of the shuttle *Discovery* astronauts who conducted the first Mir rendezvous five months earlier—and they would try to avoid one problem. On that earlier flight, cosmonaut Vladimir Titov had the microphone for the radio link between the shuttle and Mir. Gibson felt things got a bit too chatty between the two spacecraft. "There was an awful lot of visiting going on," he says. "They [*Discovery*] had Titov on board and his buddies were aboard Mir and there was a lot of reminiscing." Fortunately, the radio that Gibson used on *Atlantis* had two frequencies. Gibson would use only one, so the flip of a switch in the cockpit gave him the option to keep any unnecessary radio traffic, on the second channel, out of his ears.

The linkup between *Atlantis* and Mir would be made possible by some extra equipment in the shuttle's open payload bay. The $100 million Docking Adapter that would lock the shuttle and station together was built under a contract between NASA, the Rockwell Corporation, and RSC Energia in Russia. The one-and-a-half-ton structure was installed in *Atlantis*'s cargo bay as an extension to the shuttle's airlock that leads into the crew cabin. On top of the fifteen-foot-tall tunnel was the Russian-built docking ring, which was a copy of the one on Mir. During the linkup, the ring on *Atlantis* would ease up against the ring on Mir, triggering hooks, latches, and springs that would cushion the approaching shuttle and then draw the two spacecraft together to create an airtight seal.[23] The same docking maneuver would be used once the orbiters started work on the International Space Station. NASA was finally building skills it would directly utilize on its mission of constructing the new outpost.

After *Atlantis* and Mir docked, the shuttle crew would float over to the Russian complex by threading their way through the now connected tunnel that ended with a round hatch separating the U.S. spacecraft and the Russian station. There would be an identical hatch on the other side that the Mir crew would have to open. Before cranking open their door, the astronauts would be able to look through a small round porthole. There would be a cosmonaut staring at them from the other side.[24]

The view during *Atlantis*'s docking maneuver was different from *Discovery*'s back in February. That first rendezvous had occurred on the "nighttime" side of the orbit, when the earth blocked the sun. Everything was in darkness except for the rhythmic blinking of the beacon light on Mir. *Discovery* also maneuvered up in front of Mir for its approach. By contrast, "Hoot" Gibson and *Atlantis* would come up from below, and on the daylight side of the orbit. Video beamed to ground controllers from Mir thus showed *Atlantis* fully lit, with the continents and oceans of Earth behind it.[25] Gibson's attention was focused on a camera view from the shuttle's docking hatch, which was focused on Mir's hatch as the two craft inched closer and closer during the approach. A black metallic X on the center of the hatch was the target the shuttle crew was aiming for. In this case, X did mark the spot. Another camera view from the side documented the two docking rings, mounted on heavy springs, coming closer and closer until they touched and locked.

"Houston, *Atlantis*, we have capture," said Gibson.

"Copy, capture," Moscow replied.

As the hatches were opened for the first visit, the *Atlantis* crew were met by a whiff of cold and dank air from Mir's docking module. The astronauts later compared it to being in someone's wine cellar.[26] One thing on Norman Thagard's mind when *Atlantis* arrived was that the shuttle could solve his ongoing food problems. "I went aboard the shuttle to graze," he says. Not that the surroundings inside the shuttle made a comfortable place in which to grab a snack. "In fact, when the space shuttle docked to Mir, I preferred the Mir environment to the shuttle, because the shuttle was so sparkling clean with its white walls," says Thagard. "It was like a hospital room, which I didn't find all that comfortable after being in a nice cozy space station."

The two cosmonauts in *Atlantis*'s crew would take over for Thagard, Strekalov, and Dezhurov when the shuttle departed to return them to Earth. The handover—the process of transferring control of Mir—would require a series of meetings. For one thing, the outgoing crew had to let the incoming crew know where things were aboard Mir. They also had to change out the custom-made seat cushions in the Soyuz, so the new station crew would be able to make an emergency departure at any time after the shuttle left.

When Thagard removed his Soyuz seat, he removed as well the last glimmer of hope that his Mir mission might end the way he would have liked. He wanted to go home not aboard space shuttle *Atlantis* but rather inside the Soyuz TM. Thagard was the first American to fly up in a Soyuz, but no astronaut would land in one until the spring of 2003 when, after the *Columbia* disaster grounded the shuttle fleet, the two astronauts then aboard the International Space Station were forced to return to Earth in a Russian craft.

Thagard, Strekalov, and Dezhurov would make the trip back to Earth in reclining flight couches on the middeck of *Atlantis*. After four months in orbit, NASA was concerned that the men would faint if they were sitting upright when gravity kicked in during landing. To avoid the problem, the returning Mir crewmen would land lying down, while the rest of the *Atlantis* crew used traditional seats. For Thagard, the return to normal gravity was manageable. Astronauts usually experience disorientation as they recover their sense of balance. Often, bowing the head will give them the feeling that they are tumbling forward. Thagard found this condition worse after his long Mir mission than after shorter shuttle flights.

Other postlanding precautions, he felt, were overblown. "We've heard that if you take a hot shower, you can get lightheaded and pass out. There are even signs in the showers at the crew quarters at KSC that warn against using hot water. I took a hot shower almost immediately after landing and felt fine."

The successful return meant NASA had crossed a hurdle in its quest to build the new International Space Station. It had proved that the shuttle could dock to a space station, deliver supplies and transfer crew members, and then undock successfully.

Some heavy lifting came next.

Two years before *Atlantis*'s docking mission in 1995, NASA had voiced concern that a shuttle coming in for docking would come perilously close to bumping the "forest" of solar power panels sticking out every which way on Mir. Both countries pressed ahead with the schedule of joint missions while working to solve this problem. The solution, during NASA's second Mir docking, would give the agency the experience it needed for the first construction flight of the International Space Station. That first ISS flight required a shuttle crew to meet up with the first Russian-built station compartment, already floating unmanned in orbit. The astronauts had to connect a U.S. module to its Russian counterpart to form the nucleus of the new outpost.[27]

Astronaut Ken Cameron led the mission of *Atlantis* on STS-74 in November of 1995. This was the same shuttle that docked to Mir in July, but with a different crew and objectives. Cameron was joined by veteran astronauts Jerry Ross, Bill McArthur, and Jim Halsell and newcomer Chris Hadfield of the Canadian Space Agency. Their cargo was a docking module for Mir, resembling a large orange-colored beer keg.

The new module was designed to lengthen Mir's current docking compartment so future shuttles would link up a little farther away from the outpost's delicate solar panels. The four-and-a-half-ton compartment was equipped with docking rings at both ends of its fifteen-foot length. After its trip to space in the shuttle's cargo bay, the astronauts grabbed the module with the shuttle's spindly robotic arm.[28]

Even in the weightlessness of space, the fifty-foot arm didn't have the strength to hoist the module and snap it onto Mir. Instead the crew let the shuttle's inertia do the work. They moved the new compartment high over the docking tunnel on *Atlantis* that had made contact with Mir during

Figure 32. Space shuttle *Discovery* lifting off "early" on a mission to the International Space Station. There's a target time for launches, but during the last docking mission to Mir, NASA launched five minutes early to avoid the bad weather that was approaching. That same opportunity was used on the *Discovery* flight depicted here. Courtesy of NASA.

July's linkup in space. Then they carefully positioned the barrel-shaped module's lower docking ring just inches above the shuttle's ring. The arm was commanded to go limp, and commander Ken Cameron fired *Atlantis*'s jet thrusters so the spacecraft and the module bumped together and locked. A tower now extended up and out of the payload bay, and the shuttle was ready to dock to Mir.

The final approach followed much the same path as the previous docking. The difference was that Mission Control in Houston lost the video downlink just moments before contact and capture. Computer-generated animation took the place of Mir and *Atlantis* during confirmation of docking. When the video signal was reacquired, Mir had its new docking module, which was easily distinguished by its bright orange color. During the later undocking, *Atlantis* unlocked only the lower set of hooks and latches that connected the shuttle to the new module. The astronauts then departed for the trip back to Earth, leaving the docking module permanently attached to Mir.[29]

The scene would be replayed in 1998 when space shuttle *Endeavour* carried up the U.S.-built Unity module and used a similar maneuver to attach it to the Russian compartment called Zarya to begin building the International Space Station.

After the first long-duration Mir mission featuring astronaut Norman Thagard and two cosmonaut crewmates, more joint flights followed. Astronauts Shannon Lucid and John Blaha made long-term stays aboard Mir, as did Jerry Linenger. NASA was gaining experience in resupplying a space station by delivering water in clear plastic bags. Other items like food and clothing were stuffed into identical tan duffel bags. The astronauts kept track of the similar-looking duffels by marking them with bar codes and scanning them as they were unloaded, like cashiers at a supermarket checkout line.[30]

NASA would gain more harrowing experience during Jerry Linenger's extended mission aboard Mir. It was then that the dangers of long-term spaceflight became more apparent.

Open the Hatch Carefully

Linenger was, at that point, the least experienced astronaut to make a long-duration flight on Mir. Each of the people who preceded him on the outpost had logged four shuttle missions. Linenger had only one before being named to a Mir flight.

His smile beamed from the crew photo of STS-64, a *Discovery* mission featuring the release of a solar satellite and a spacewalk where no safety lines were used. Spacewalkers usually have a tether to keep them from floating away. On this mission, NASA wanted to test an emergency jet pack attached to the spacewalkers' backpacks. The crew members successfully jetted their way around the cargo bay during the test flights.

Linenger's next mission was the long one aboard Mir. He trained to fly a Russian Soyuz space capsule, like Norman Thagard. But that would be for an emergency departure from Mir and a landing back on Earth. Linenger would fly to the Russian space station aboard space shuttle *Atlantis* on STS-81. Linenger's crewmates for his long-duration flight, cosmonauts Valery Korzun and Alexander Kaleri, would already be aboard Mir, having arrived on an earlier Soyuz launch from Russia. As *Atlantis* closed in for docking, both Linenger and crewmate John Grunsfeld marveled at the collection of crew modules clustered together and solar panels sticking out randomly. Their opinion changed, however, once the shuttle docked firmly to the aging outpost and the hatches were opened.

"It smelled like your grandmother's attic," says Linenger. "And you had to push stuff out of the way to move around inside."

It became clear that the image of the Mir station he had been carefully fed by his instructors at Star City in Russia was a flattering one. For example, the Kvant 1 module attached to the rear of the main crew module was advertised as an advanced astrophysics module. "That may have been the case fifteen years before I got there," says Linenger. "But what I saw was an old rusting compartment, stuffed with gear like a storage room." Along with the smell, Mir was cluttered and dark, making even the simplest trip around the station like exploring a cave. Under normal operation, the cosmonauts conserved power aboard Mir by keeping the lights low. That is, until the press was invited to ask questions during televised news conferences. Then, Linenger noticed, the lights were brought up full, and wide-angle camera lenses were used, giving visitors on Earth the impression that the Russian station was spacious and vibrant. "The Russians wanted to put their best foot forward," says Linenger. "They wanted to sell the station to the Americans, and they needed NASA's money and backing."

After the period of adjustment, the first weeks aboard the station went fine and Linenger made himself at home. "I basically picked a spot in one module and made a spot on the wall for me to sleep," he says. "That was just my spot." The departure of the shuttle was a lonely moment for the sole astronaut left aboard Mir. These early docking missions were preparation for the International Space Station, and a way to breathe life back into the shuttle program. Still, the winged vehicle felt like Linenger's lifeboat, and that ship had just sailed without him.[31]

One surprise for the astronaut was the sense of isolation as he worked in Mir's various compartments. He didn't feel cooped up with strangers, he says, nor did he feel like he was in solitary confinement. "I'd be working in one compartment and the other cosmonauts would be somewhere else. And I'd float into one cabin and pass Vasily and say *zdravstvuite* [hello]. I didn't feel the need to get away from anyone."

One thing Linenger remembers of his time aboard Mir is the noise. "The air didn't circulate, so there was the constant humming of fans everywhere you went." That made the frequent power failures something the astronaut looked forward to. The station grew dark, but the whirring of the fans mercifully stopped for a time. Technical glitches, however, started piling up, and the trend didn't escape Linenger's notice. Mir seemed like a 1960s car that was desperately in need of repairs that the owner couldn't afford. It also reminded him of a phrase he heard again and again while training in Russia to do spacewalks:

Open the hatch carefully.

NASA and Moscow used a similar technique to train crew members to work outside a spacecraft or station in heavy spacesuits. They put their spacewalking astronauts and cosmonauts into a deep pool of water to mimic the effect of floating in weightlessness. Linenger underwent this kind of practice in a Russian-made spacesuit called an Orlan. As he rehearsed the work he would eventually perform aboard Mir, his teachers would whisper to him through the headset he wore inside his helmet. "Open the hatch carefully" was the message. Linenger quickly learned why that was important once he was performing his first real spacewalk aboard Mir while orbiting the earth at 17,000 miles per hour. "The hatch from the airlock to the outside was held in place with C clamps," he says. "Kind of like the ones you'd buy at the hardware store." It turned out the hatch hinges had been damaged during previous spacewalks and the clamps were put on to hold it together. That's why mission managers considered

opening the hatch carefully an important thing to remember. Also, going from the filtered air supply of his spacesuit back to the dank-smelling
atmosphere inside the space station reminded him that the outpost had
seen better days. "My flight marked a downturn aboard Mir," recalls the
astronaut. "Part of the time we were working to keep the station alive, and
part of the time to keep ourselves alive."[32]

The Fire

Matters got worse after another Soyuz craft docked to the Mir station on
February 12. The crew included two cosmonauts and a guest researcher
from Germany. Eleven days later, fire broke out.

The crew on Mir gets oxygen to breathe in one of several ways. There's
a device called Elektron that separates the oxygen from the hydrogen in
urine from the crew. Another source is called an oxygen canister. The
crew lights these like candles, and the burning process creates breath-
able oxygen. That was the technique of choice when Linenger was in the
Kvant 1 module attached to the main crew cabin. A master alarm sounded,
and a crew member floated past trailing smoke. "One quick flyover," says
Linenger, "and you could see a three-foot flame coming out like a blowtorch."

A fire in any inhabited structure can be life threatening. In space the
danger is worse, because fires in weightlessness don't behave like flames
on Earth. In a house blaze, a basic safety rule is to get down low: smoke
rises, and you can often find air when you get on your knees as you crawl
to safety. That bit of logic was of no use to Jerry Linenger. In space, smoke
spreads uniformly, and you can't just open a window to clear it out. He
needed one of the oxygen masks on Mir and he needed it fast. "It's similar
to swimming twenty-five meters underwater," says Linenger. "The smoke
was very dense and you couldn't take a breath." He found a mask, strapped
it on, turned a valve, took a deep breath, and got nothing. The mask wasn't
working. "At that point, I threw the mask aside and tried to grab a breath
down low. Earth logic didn't work up in space. I grabbed a second mask,
flipped it on, and must have hyperventilated for thirty seconds."[33]

Linenger's immediate needs were met. He had air. But the danger was
far from over. The fire was still burning, and the alternatives were bleak.
He and commander Valery Korzun set about fighting the fire while cos-

monaut Kaleri prepared one of the Soyuz capsules for immediate departure in case the blaze grew out of control.

Abandoning ship, however, was not as attractive an option as it may
have seemed. There were two Soyuz capsules docked to Mir, but one was
at the back near the fire, with the blaze blocking the way, and the remaining Soyuz could hold only three people. The fire had to be controlled or
people might have to be left behind. "I told myself, we're going to fight this
fire," Linenger recalls, "and we're going to do everything right and we're
going to get out of this."

Even after the blaze aboard Mir was put out, the crew members faced
danger. "After the fire, it wasn't fine," says Linenger. "We still had a lot of
smoke. It was a race for the oxygen masks to last until the smoke had a
chance to clear." After about an hour the air supply slowly began to return
to normal and the masks could come off. Again, the crew didn't have a
chance to relax. Linenger was a trained physician whose job it was to set
up a CPR station and check his crewmates for smoke-related injuries. No
one was seriously hurt, and Linenger attributes that to the thickness of the
smoke that filled Mir's cabins. The six men couldn't breathe under those
conditions, so no smoke inhalation occurred.

The Good Memories

With all the danger and the disappointments, Jerry Linenger still says
he enjoyed his adventure aboard Mir and respects what his Russian colleagues had accomplished. He compares his time on the station to being
an explorer like Lewis and Clark, with trips by canoe that may not be
comfortable or totally safe but are very worthwhile. That pioneer image
was reinforced by the *Atlantis* astronauts who eventually came to pick
him up when his tour aboard Mir was completed. The crew included Jean-
François Clervoy, representing the European Space Agency, and shuttle pilot Eileen Collins. Linenger's hair had gotten a bit shaggy over the
months, and Clervoy offered to give him a trim. "Jean-François said he
wanted to make me look presentable for my wife before we returned to
Earth," says Linenger. "Then Eileen said I looked like Robinson Crusoe,
and I told Jean-François to leave my hair as it was." He liked the pioneer
image after all.[34]

Lessons Learned in Space and in the White House

Despite periods of tension, confusion, and even danger, White House science advisor John Gibbons remained upbeat as work progressed toward the joint International Space Station. The point was that astronauts and cosmonauts were working together to solve problems. "We learned a bushel of lessons over how to make this thing [the new station] more resilient and more reliable," says Gibbons. "Even fire detectors and things like that." Still, inside the White House, he says, it was becoming apparent that Mir was being held together with duct tape and string.

There would be more bad news for Gibbons to deliver to the President after astronaut Mike Foale settled in aboard Mir. He would have a close call of his own while one of the cosmonauts was using remote control to pilot an unmanned Russian cargo ship called a Progress. The craft went out of control, crashed into the Spektr science module, and punctured it. Foale and his two crewmates could feel the loss of air pressure in their ears as their precious oxygen supply seeped out into space. The Spektr compartment was sealed off for good, along with the equipment and personal property of the crew members that was inside. The growing list of problems meant longer and longer hours for John Gibbons in his job of keeping Mr. Clinton up to date. "I basically had to give the president almost daily reports on the situation on Mir," says Gibbons. "It got to be almost hilarious. Every day things went wrong, but things got resolved or fixed. So the history of that back-and-forth is the ingenuity of the Russians and Americans to make that thing [Mir] work until the end." As NASA's joint missions with Russia aboard Mir drew to a close, both agencies looked ahead to the problems and challenges facing the International Space Station and the mixed support the project was getting in the U.S. Congress.

The station would be jointly built and controlled with the Russians, with American taxpayers paying much of the cost. If anyone knew what this arrangement would be like, it was the astronauts who worked, and sometimes fought for their lives, aboard Mir.

The Russian station would finally be abandoned and later plunge, unmanned, like a fireball into the Pacific as Russian ground controllers guided the complex through a controlled burn-up in the atmosphere. The next step was building the International Space Station, which would finally be a mission the shuttle program could call its own. From the perspective of the White House, it was also a chance to deal with the final "enchilada."[35]

9

1998

ISS, the Shotgun Marriage

Finally, the space shuttle program was about to embark on a true mission. After nearly two decades of mundane flights to launch satellites and conduct endless experiments, NASA believed that building the International Space Station would truly utilize the potential of its troubled shuttle. It would also require a steady stream of launches to build skills and confidence along the way, like the Apollo moon flights in the 1960s.

During Project Apollo, which had a specific mission to achieve, each consecutive rocket launch developed specific abilities the astronauts would need to eventually allow Armstrong and Aldrin to walk upon the moon during the flight of *Apollo 11.*

The first of this series, *Apollo 7* in 1968, tested the mother ship, made up of command and service modules, in orbit around the earth. Later that year, the *Apollo 8* astronauts flew to the moon and took the famous "Earthrise" photograph of our planet rising over the barren lunar surface. That flight proved that people could go to the moon and return safely. In early 1969, *Apollo 9* tested the buglike lunar module that the first landing crew would utilize for touchdown. With all that preliminary work done, *Apollo 11* was ready to put humans on the lunar surface.

In this same way, NASA was learning step by step how to build and operate a space station by working with the Russian space station Mir. Now, in 1998, both nations were ready to move on to constructing the International Space Station. The Russian sections would be launched on unmanned rockets, just as the components of the defunct Mir station were. All of the parts from the United States would go up on the shuttle, and as these parts were being installed, spacewalkers would work outside to extend connector struts and plug in cables and fluid lines. NASA would finally have the opportunity to demonstrate what human capability on its reusable spacecraft could achieve during a long-term project, assuming everything went well in Congress and on the launchpad.

When the U.S. and Russia started talking about building a space station together, it was clear Washington had money that Moscow lacked and Moscow had experience and equipment that NASA lacked. NASA also had the ability to carry up heavy cargoes with the space shuttle, although for the Russians that was a mixed blessing due to the risks of the space shuttle made clear by the *Challenger* accident. "Russians have had their share of space disasters," Sagdeev reminds us. "The first pilot of the Soyuz was killed, and three men who went to the Salyut 1 station died when pressure was lost in their return capsule." There was also a lot of debate in the Russian scientific community over whether the basic design of the shuttle was the smart way to go, and whether it was worth the money. "The decision was that it was too complicated and expensive," says Sagdeev. "And I was on the skeptical side." Despite the divisions among Moscow's leading scientists, politicians in the Soviet Politburo voted in favor of building a Soviet version of the shuttle, to be called Buran. It flew once, unmanned in 1988.[1]

As far as teaming up the shuttle and Mir was concerned, U.S. mission planners and congressional budget officials were discussing whether the Russians could deliver on the parts of the station they promised. Lawmakers on Capitol Hill also balked at the cost, much of which would be shouldered by U.S. taxpayers.

NASA already had its shuttle fleet flying. So, mission or not, the primary expenses were for day-to-day maintenance, training the crews, and launching the spacecraft and whatever payloads had to be built. By 1995, contractors like Boeing were busy building the basic modules for the station, like the Destiny science lab and the connector node called Unity, which would act as a hub for bigger U.S. and Russian compartments. That meant lawmakers on Capitol Hill who were critical of the whole station project had time to complain, and perhaps kill it.

The one legislator who became emblematic of Congress's disdain for the International Space Station was Indiana Democrat Tim Roemer. When the House came within an eyelash of voting to cancel the project in 1993, Roemer was credited with leading the opposition. Supporters barely garnered enough support to keep the station going. "Save the space station!" cried Democratic congressman James Bacchus of Central Florida from the podium during a last-minute appeal. Many space-related jobs in his district depended on the outcome, and the project survived the axe by a single vote. That fight wasn't the first one over the station, and it wouldn't be the last.[2]

NASA's budget falls largely under the control of the Science Committee of the U.S. House of Representatives. It was there that Congressman Roemer first developed his dislike for the space station. "It was back in 1991 when I had my first hearing on and heard the promises about this technology," says Roemer, "and how it would abide by the 1984 budgetary goal of eight billion dollars in cost and return great science." Instead, Roemer and critics of the station complained, it produced billions of dollars in cost overruns and accounting practices later described as Enron-like. Even supporters like White House science advisor John Gibbons had to admit that, despite the best of intentions, the project was getting more expensive. As NASA approached the early stages of construction, Roemer didn't let up on his criticism. "I don't think they [NASA] will be able to afford or have the credibility to build the kind of station that will do life science and materials science research," he said, "and it will be generally a big waste of money." Those complaints were echoed by Representative Dana Rohrbacher, chairman of the House Subcommittee on Space and Aeronautics. The argument centered on escalating costs, allegations of poor management by NASA, and the probability of low scientific return on the investment. During a hearing before his subcommittee on November 5, 1997, Rohrbacher made clear his concerns over the growing cost overruns associated with the station. The first pieces of the orbiting outpost would not be launched for a year, and the price kept going up and up. "So what's going on with the space station?" Rohrbacher asked. "What is it with these big hardware projects that became juggernauts that no one seems to be able to manage?"

Judging from the numbers, Roemer and Rohrbacher and the other station critics had reason for concern. The cost was originally supposed to be eight billion dollars. But as the launch date for the station drew closer and closer, Congress pointed out, NASA was transferring money from developing the science research that was supposed to be done aboard the station to extra expenses related to basic construction. Some $220 million was shuffled from science into hardware in fiscal year 1996, with additional transfers in 1997 and 1998. In his comments before the Science Subcommittee, Rohrbacher complained that NASA was looking at a possible billion dollars in extra expense for the station. "Frankly," said Rohrbacher, "the only technological breakthrough this program has managed is to magically turn my friend Tim Roemer from a gadfly into a prophet."[3] Critics of the station often pointed to a GAO report that, at one time, estimated the full cost to build and operate the complex during its lifetime at 100 billion dollars.[4]

The station project, which held the key to an actual mission for the space shuttle, was clearly in peril of not getting off the ground. The White House weathered the storm in Congress and argued that the outpost was a good way to develop international friendships while working on a project in a nonwarlike manner. But still there was no hardware in space, and that had to change.

The new International Space Station was a shadow of what space station Freedom was supposed to be, but it was a shadow Congress appeared somewhat comfortable paying for. Originally, Freedom was to have had a huge square-shaped truss of girders surrounding the crew compartments and forming the spine of the station. The proposed Freedom station would be powered by thermodynamic solar generators. The new technology would use huge mirrors to track the sun and focus solar energy to create electricity.[5]

Gibbons's primary complaint about the older Freedom station idea was that it was a relic of the Cold War. Back then, Freedom was to be made up of mostly American modules, and it had a decidedly "American" orbit—something that would be changed for the new station called

Figure 33. Artist's conception of NASA's space station Freedom. The orbiting complex never flew, but later evolved into the International Space Station, which features Russian participation as well as resources from Europe, Japan, and Canada. Courtesy of NASA.

ISS. "We put the orbit at a higher inclination, so it could pass over both Russian launch sites and American ones," Gibbons recalls. "We avoided that before [with Freedom] when it was a Cold War project. Now, that meant the Russian facilities would be serviceable to it." Orbital inclination is the angle at which a spacecraft or space station crosses the equator. The shuttle frequently takes a 28-degree inclination, hugging the equator as it circles the earth. Steeper angles like 50 degrees make the vehicle take a roller-coaster-style path farther north and farther south. The new space station would have a 51-degree inclination, which would make it pass within range of spacecraft launched from Russian pads as well as from the Kennedy Space Center.[6] That proved to be a godsend following the *Columbia* disaster in 2003, which grounded NASA's space shuttle fleet. For more than two years, the only vehicles able to reach the orbiting outpost were Russian-built Soyuz crew capsules and robotic Progress cargo ships

launched from the Baikonur Cosmodrome in the former Soviet nation of
Kazakhstan.

During the life of the Space Station Freedom program, token represen-
tation by foreign nations had been mandated by the Reagan administration. Participants like Europe, Canada, and Japan were known derisively
as "junior partners." Now, having the Russians on board was considered
critical.

The new station would have a smaller truss shaped like a capital letter I, with the pressurized crew modules clustered at the center. Japan
signed on to provide a science compartment called Kibo, meaning "hope"
in Japanese. Researchers would float in and out of the crew module while
a mechanical arm moved science experiments outside on a "porch" open
to the heat and cold of space. The European Space Agency would build its
own laboratory, named Columbus, as well as an observation dome called a

Figure 34. The Japanese experiment module, known as Kibo. Designed and built
by the Japanese Space Agency, JAXA, for attachment to the International Space
Station, the pressurized laboratory module also features an experiment "porch"
where a robotic arm moves science packages in the vacuum of space. Photo by Pat
Duggins.

cupola. The scaled-down station would have no solar generators, just solar panels the length of a 747 passenger jet's wings.[7]

All in all, the International Space Station was smaller than the proposed Freedom station, and the new outpost was half Russian. NASA's "cabin in the sky" was now a duplex. But half a station was better than none.

East Meets West to Build the First Sections of the International Space Station

The first two parts of the station went by the nicknames Unity and Zarya (meaning "dawn"). These were far more aesthetic than the technical names Node 1 and Functional Energy Block.

Zarya would be the first Russian section of the International Space Station. That is, unless you asked Boeing. In that company's opinion, the Zarya compartment was only partly Russian. It would contain U.S.-made computers and power converters to adjust electricity from Russian solar energy panels so the American sections could use it. Zarya would also return the favor, so the energy from U.S. solar panels was the right voltage for the Russian side of the multinational station.

The compartment, known as the functional energy block or FGB in Russian, would be the size of a railroad car and resemble the older Mir space station's Krystal module that shuttle commanders used during delicate docking maneuvers.[8] Boeing was selected by NASA to subcontract the building of Zarya from the Krunichev factory in Russia, which built the modules for Mir. For the U.S. contractor in charge, it was a unique window into how the Russian space program worked. "They have a thirty-year legacy of successful space work," said Virginia Barnes, FGB program manager for Boeing. "One of the lessons they learned is, if it's not broken, leave it alone."[9] That became apparent in the size, shape, and configuration of Zarya. It closely resembled the previous Mir compartments, which evolved from the older Salyut space stations. The Mir modules would be the template for Zarya and the upcoming crew cabin for the International Space Station, which would be known as Zvezda, or "star."

The plan was for the Russians to launch the Zarya compartment on an unmanned Proton rocket, the same way the former Soviet Union launched Mir. The U.S. space shuttle would carry up the Unity node and snap the two parts together in space. Boeing's Virginia Barnes was in charge of ac-

Figure 35. Artist's conception showing space shuttle *Endeavour* just moments before connecting the first two pieces of the International Space Station. The astronauts grappled the U.S.-built Unity module and snapped it onto the shuttle's docking tunnel. Step two was to grab the Russian-built Zarya compartment, position it just above Unity, and then fire the shuttle's jets to push the two halves together. Courtesy of NASA.

quiring Zarya from the Russians, and the East-meets-West scenario was unsettling at first. "It's like anybody would be, on their first trip to Moscow," said Barnes. "You might notice differences that are strange at first, like there's Cyrillic [Slavic alphabet] on the signs and so forth. It gives you the feeling that you know what it's like to be illiterate." These Boeing workers had to ride herd on the Russian technicians, many of whom felt they helped make Moscow the world's leader in space station technology and experience, thank you very much.

The adjustment for both sides was jarring. Boeing workers, who were used to reams of documentation on each piece of space hardware to certify that it was ready for flight, would tell stories of walking into the workshops at Krunichev and finding components scattered around and almost no paperwork at all. "I wouldn't call them horror stories," said Barnes. "There are just a lot of differences in the Russian approach to manufacturing and testing. What might have been construed as mass confusion really wasn't. Initially, there was some reluctance on the Russian side to share data. It was very important for the Russians to know why you needed the data. If you said why, you were likely to get it." The twenty-ton Zarya underwent extensive testing to ensure it would do the job all sides needed once it was in orbit. The compartment would be the connector between the Russian modules and the American side. There would be electricity fed through Zarya, as well as computer data, fuel storage, life support, and thrusters to keep the station flying straight. Zarya would have solar power panels and docking ports to receive Soyuz crew capsules and unmanned Progress cargo vehicles. Boeing would clearly have to calm down and allow the Russians to be themselves during the construction and launch process. "Paint and polish and spit shine is important to us," said Barnes. "They have a different way of getting things done."[10]

Where Zarya looked like part of a space station, Node 1, or Unity, seemed less impressive. It looked like a barrel.

The node was what the name suggested, a connector. The squat silver compartment had docking hatches on top and bottom, front and back, and left and right. At the front and back of Unity were short, conical black tunnels with docking adapters, covered with electrical cables. Those tunnels were called Pressurized Mating Adapters, or PMAs. Zarya would snap onto the rear PMA, and the space shuttle would link to the front one. Designers compared the Unity node to the center of a small town. It didn't seem to have a particularly exciting function, except to link together other modules that did.[11] Processing Unity for flight aboard *Endeavour* was the job of Roy Tharpe, head of Boeing's Florida operation. "On Unity, you can go in the front door, or out the back door," he said. "Or you can go into the kitchen, or down into the basement." More specifically, crew members could float back through the Unity module into Zarya, forward into a U.S. laboratory module called Destiny, to the left into an airlock called Quest, to the right into a small observation dome called a cupola, or down into a planned habitation module. From NASA's perspective, Unity might not have been as sexy as the Russian modules, but at least it was ready to go to space.

Launch delays kept Zarya on the ground seventeen months longer than expected, and the snags infuriated NASA. The U.S. agency needed the first two Russian modules to power the fledgling station and to provide thrusters to keep it pointed in the right direction and at the right altitude. Unity would have none of these abilities.

Moscow had to come through.

After the seventeen-month delay, as NASA prepared *Endeavour* for the launch of Unity, astronaut Jerry Ross summed up the frustration during a press conference. "It's ready to go, we're ready to go, we've been waiting a long time and we need to do this."

Ross and crewmate James Newman would float outside on spacewalks to connect power and data cables between Unity and Zarya once the two modules were locked together. Astronaut Nancy Currie would fly on *Endeavour* as well. Her job was to grab Unity with the robotic arm and perch it on the shuttle's docking tunnel in the cargo bay. *Endeavour*, with Unity sticking up and out into space, would slowly close in on Zarya. The Russian module was supposed to be waiting in orbit following its carefully coordinated launch on an unmanned Russian rocket. Currie would grasp Zarya, position it above Unity, line up the docking adapters, and, with a nudge from the shuttle's thrusters, bump and lock the two compartments together. The trick now was to get Unity safely into orbit.[12]

The process of loading Unity for launch occurred with considerable ceremony at the Kennedy Space Center.

Reporters were herded through the labyrinthine space station processing facility near KSC headquarters. The path took the press over a large sticky welcome mat designed to pull dust and dirt off the visitors' shoes. Then it was up a metal staircase and onto an observation platform over the large U.S.-built pieces of the International Space Station.

Technicians in blue coveralls called "bunny suits" worked intently on the solar panels and other station components waiting for a ride to space on the shuttle. Anyone coming within three feet of the flight hardware had to wear a bunny suit, complete with hair cap, bootees, and a mask if the wearer sported a beard. The concern was over something even as small as a fleck of dust coming off someone's clothes and jamming a valve on a station module.[13]

Soon a crane lifted the Unity module from its cargo stand on the floor of the large room, up over the heads of the reporters, and slowly toward a waiting cargo canister that would carry the node to the launchpad to be packed aboard *Endeavour*. The shuttles had been flying since 1981 and looked that way. Millions of miles in orbit and dozens of space missions left the space planes scuffed and worn. Unity had that showroom look of something brand-new and fresh from the factory. Just feet over the heads of the onlookers, the node's silver aluminum hull could be seen marked with white dots that future shuttles would use to target the docking tunnel on final approach. Where the cone-shaped black docking tunnels attached to Unity at the front and back, each of the stout adapters angled down from the compartment, giving the whole unit a slightly off-kilter appearance. NASA's Glenn Snyder called them "black joggles."

Snyder was payload manager of STS-88, *Endeavour*'s mission to deliver Unity. He seemed amused by the notion of calling the U.S. payload by its nickname instead of its "real" name of Node 1. "We liked the term 'node' because we're engineers. When we looked at schematics, we located node 1, node 2, node 3, and so on." Node 1 was also a good enough name for the road signs on the outside of the module. "Node 1" was stenciled on the yellow handrails along the compartment's surface. The point was to guide spacewalkers as they moved along the exterior of the outpost. "They [NASA] found that when we put up acres and acres of space station, it would be tough for spacewalkers to know where they were, so we put signs on the handrails saying 'node 1' and so on. So if Mission Control says, 'Go to node 2,' they'll make their way along these labels until they find node 2."

The name Unity was clearly for public relations, just for show. The tension on launch day probably wasn't helped by the presence of Secretary of State Madeleine Albright and her entourage of U.S. Secret Service agents.

Endeavour Launches the First U.S.-Built Chunk of the International Space Station

Endeavour commander Bob Cabana and his five crewmates were dealing with the usual prelaunch concerns while strapped in the shuttle's cockpit and middeck. Rick Sturckow was making his first flight. "If there's ever a fear of flying the shuttle, it's when you're at the one-hundred-and-eighty-five-foot level," that is, on the gantry before boarding the spacecraft, Ca-

Figure 36. The crew of mission STS-88 on the way to the launchpad to carry up the first U.S.-built segment of the International Space Station. First row, left to right: astronauts Rick Sturckow, Nancy Currie, and Bob Cabana. Second row, left to right: cosmonaut Sergei Krikalev and astronauts Jerry Ross and Jim Newman. Courtesy of NASA.

bana admits. "That's when you come to grips with your mortality." One way the *Endeavour* crew tried to make light of the pressures and dangers of the mission was by dubbing itself Dog Crew 3—a term that requires explanation.

The tradition began with a military mission aboard shuttle *Discovery* in 1990 with a crew calling itself the Dogs of War. The nickname stemmed from commander Dave Walker's thatch of red hair, which earned him the name Red Dog during his days as a navy fighter pilot. Each astronaut took a nickname that fit into the dog theme during that flight, and on the next mission that Walker commanded as well. *Endeavour's* crew claimed that theme as part of the first mission of the International Space Station.

Bob Cabana went by the name Mighty Dog. Former marine fighter pilot Rick Sturckow was known as Devil Dog, spacewalkers Jerry Ross and Jim Newman were Hooch and Pluto respectively, and Nancy Currie went by Laika in honor of the first canine to go into space on a one-way trip for the Soviet Union. Cosmonaut Sergei Krikalev might get the highest marks for creativity with his Russian-inspired dog name of Spotnik.[14]

Latching the two modules together was only job number one. The hatches into the outpost had to stay locked shut while spacewalkers Jerry Ross and Jim Newman prepared for the first two work sessions outside the complex. Forty electricity and data cables had to be connected so energy from Zarya's solar power panels could flow to the American node. The protective cover of each cable had to be removed and each plug inspected to make sure there were no bent pins. Then each line coming from Unity had to be connected to a specific mate coming from Zarya and a latch thrown to lock the two ends together.

Each connection that was successfully finished made U.S. contractors feel a little better for insisting on all that paperwork from the Russians. The plugs from Unity had to fit just right into their counterparts built and installed on the Russian module half a world away. "You want to make sure oxygen lines connect to oxygen lines and not nitrogen lines," says STS-88 payload manager Glenn Snyder. "That's why we asked for the documentation, so if there's a problem we can unravel it." Just to make sure, though, Snyder says NASA physically took each connector plug to Russia and practiced sticking it into the receptor on the Zarya side.

Spacewalk number two outside *Endeavour* was to install a pair of ninety-pound radio antennae to link the station to ground controllers both

Figure 37. The STS-88 crew posing for the traditional astronauts' breakfast photo prior to suit-up and departure for the launchpad. The cake before them is decorated with the crew patch designed by the astronauts for their flight. Left to right: cosmonaut Sergei Krikalev and astronauts Rick Sturckow, Nancy Currie, Bob Cabana, Jerry Ross, and Jim Newman. Courtesy of NASA.

in Houston and in Moscow. Jerry Ross also worked to loosen a backup antenna on Zarya that failed to pop open following its launch.

On December 10 the space station was opened up for the first time. Cabana and Sergei Krikalev moved from *Endeavour*'s airlock and into the black docking cone coming from Unity. A crank was turned on the large square hatch separating the shuttle crew from the station's interior. The door slid open, and the astronaut and cosmonaut began to encounter something rare for people in space—elbow room. The box-shaped compartment was spacious, allowing crew members to float freely from one white wall to another. "Going into the International Space Station, what a thrill," Bob Cabana recalls. "Just the anticipation of turning on the lights and getting it ready for future missions, you could see it in the faces of the crew." The shuttle crew later opened the hatches between Unity and Zarya. The Russia compartment was even roomier than Unity, with a long hallway framed by clean white walls lined with handrails.

The five astronauts and one cosmonaut went about the job of installing a teleconferencing system and removing protective bolts from the interior of Zarya that were designed to keep the panels on its walls from popping open during launch. Cabana later took a video camera with a long cord and threaded his way from *Endeavour* into Unity, past crewmates working behind panel covers there and in Zarya, down the long passageway of the Russian module and back again.[15]

The extra room had a downside for astronaut Nancy Currie, the shortest member of the six-person crew. "We were waiting inside Unity for a press conference," says Cabana. "And Jim Newman and Sergei Krikalev were kind of pranksters. And they called Nancy over and held her steady in midair in the middle of the compartment, just let her go, and left the room. She reached out for a handrail and couldn't get one. She was stuck for about five minutes until air from the air vents pushed her to the side. She's more careful around those guys now."

Still, the jokes and handshakes between American astronauts and cosmonaut Krikalev couldn't mask growing problems between the United States and Russia over the new station and clear distrust in the halls of Congress. Even as NASA was still taking bows for the successful first step in building the International Space Station, critics in Washington were mounting a new battle to end the project before any more parts were sent up and any more money was spent. A hearing was held by the full House Science Committee on May 18, 1999, to consider funding for NASA for the next three fiscal years. That included the space shuttle, the space station, and science research that would be done, and some members of the committee felt there were big problems. The spending under consideration would include roughly $2.5 billion to operate the shuttle program, with equally large pots of money for the station and space-related science. The growing price tag for the International Space Station and complications regarding Russia's participation were at the forefront of the debate. Lawmakers complained that the cost of building the space station had, at that point, risen from $17.5 billion to nearly $25 billion. Despite Russia's delivery of the new crew module for the station, members of the Science Committee still chastised Moscow for slow progress. The former Soviet Union's job was to build and launch compartments that were supposed to make up half of the outpost.[16] Republican Mark Sanford of South Carolina offered an amendment that would have a chilling effect on the station. It would cut all funding completely. Lawmakers rose in support of the idea. The name of Congressman Tim Roemer, chief critic of the station, was invoked a number of times. The list included Representative Lynn Woolsey of California, who asked to speak as "Tim Roemer in a skirt." The self-proclaimed former welfare mother who was first elected to Congress in 1993 referred to the GAO study that stated that running the space station would cost $100 billion over its lifetime.[17]

Woolsey declared that the station was casting a shadow over NASA and that the money for the complex could be spent elsewhere. "We need to shore up Social Security and Medicaid," she said. "We need to educate our children and clean up the environment. We need to get our spending priorities straight, and I support the elimination of funding for the space station." The representative listed dozens of worthy programs that could be funded, were it not for the station. The National Institutes of Health, she contended, could operate for sixteen years with that kind of money. Or the nation could fund drug prevention programs or build three new nuclear aircraft carriers. To be sure, Woolsey's district included Marin and Sonoma Counties in California, far from the nearest NASA space center, so few of her constituents would be impacted by the station's demise.

Republican David Weldon of Florida, on the other hand, had a lot to lose. His district included the Kennedy Space Center, and he leapt to the defense of the program. He pointed out that much expensive station hardware was already built and waiting to be launched, and that to stop the project now would be a waste of the twenty billion tax dollars invested to date. "We have two elements up there already," said Weldon, referring to Unity and Zarya. "And there's nothing that excites children in America more than when you talk to them about a space station and a manned space program." A vote was held to decide the fate of the proposed spending cut. As chairman James Sensenbrenner of Wisconsin asked for a recorded vote to confirm the outcome, the author of the amendment, Representative Sanford, asked to withdraw the motion so he could offer it on the House floor. The amendment came up before the full House a week later and was defeated. The space station would live to fight another day.

In the winter of 1999, the concerns voiced in Congress regarding slow progress by the Russian space program became more apparent and left Washington and Moscow on the brink of a split. Russia said the Zvezda crew module wouldn't be launched until maybe the summer of 2000. NASA administrator Dan Goldin hinted to reporters about replacing the Russian module, temporarily or otherwise, with a U.S.-built compartment called the Interim Control Module. People couldn't live inside the box-shaped ICM, but the unit could provide a guidance system to keep the

Figure 38. The STS-88 crew in front of space shuttle *Endeavour* following the successful mission to connect the U.S.-built Unity module to the Russian-built Zarya compartment. Left to right: astronauts Jerry Ross, Nancy Currie, Rick Sturckow, Bob Cabana, and Jim Newman, and cosmonaut Sergei Krikalev. Courtesy of NASA.

other two modules pointed right. The compartment also had thrusters fueled by large red spherical propellant tanks, and radio gear to keep the station operating.[18]

The continuing launch delays for the Zvezda crew cabin also meant some shuffling of the people going up on shuttle missions. *Atlantis* was scheduled to launch in May of 2000 with three crew members trained to work inside the Zvezda compartment. But Zvezda wouldn't be there when they arrived. NASA switched those people to a later flight. The two men and one woman who would be the station's second long-duration crew were put on *Atlantis* to gain experience before beginning their ISS mission.

Zvezda finally arrived in July of 2000, and that cleared the way for the first full-time crew on the outpost.

For astronauts who flew more than one flight to the International Space Station, the biggest impression was seeing how it changed over time as new pieces were added. The arrival of the new crew module meant there were three compartments orbiting the earth every ninety minutes.

The Station Comes Together, One Piece at a Time

In October of 2000, the central brace for the station's long spinelike truss framework was snapped onto the Unity module. Later that month, the first full-time crew arrived aboard a Soyuz space capsule. The next month, the first section of the outpost's 240-foot-long solar energy panels was perched in place on a long mast. Now visitors to the complex would be greeted by the sight of the long gold arrays, as wide as the wingspan of a 747.

Figure 39. The STS-104 crew of space shuttle *Atlantis* after delivering the Quest airlock to the International Space Station. Left to right: astronauts Jim Reilly, Janet Kavandi, Steve Lindsey, Charlie Hobaugh, and Mike Gernhardt. Courtesy of NASA.

Soon astronauts installed an airlock nicknamed Quest, a U.S. laboratory called Destiny, and the station's Canadian-built robotic arm. The Russian space program also added a spacecraft docking module known as Pirs. The station project clearly gave NASA a new sense of purpose, reinforcing the notion that a firm mission was the tonic the shuttle program had lacked for years. The difference was clear to see.

While astronauts flew *Discovery, Endeavour*, and *Atlantis* on visually stunning missions to the International Space Station, NASA struggled to find work for *Columbia*. The oldest member of the shuttle fleet was also the heaviest. When *Columbia* was built, it was an experimental vehicle weighing 178,000 pounds. In creating that first spacecraft, NASA gained the experience it needed to shave pounds off its later shuttles. The agency's second craft, *Challenger*, was a ton and a half lighter than *Columbia*. When *Discovery, Atlantis*, and *Endeavour* came along starting in 1983, *Columbia* was as much as three and a half tons chunkier than its younger siblings. Given the tight weight budget during liftoff, NASA would exclude its oldest shuttle from construction missions to the International Space Station. There was a limit on how much a shuttle could carry to orbit, and every ounce counted. That total included the fuel supply and the weight of the external tank and rocket boosters. After that, the leftover tonnage was split between the shuttle and whatever cargo had to be carried up, so the heavier the spacecraft, the lighter the payload had to be. The heavy parts that needed to be launched to orbit for the space station tipped the scales in favor of the lighter shuttles. They were called upon to do the work, not *Columbia*.

There would be jobs for NASA's oldest shuttle to do. Seven astronauts flew aboard *Columbia* to repair the Hubble Space Telescope. Then a microgravity science mission was scheduled with an international crew, including the first astronaut from Israel.

Columbia

The space shuttle program was hard at work on the mission to build the International Space Station, which became NASA's focal point. After the first U.S.-built module of the outpost was carried to orbit by shuttle *Endeavour* in 1998, the astronauts knew they would be doing one of two things when they were assigned to a flight. They would be called upon to tinker with the Hubble Space Telescope, or they'd visit the station. That was it—with one exception.

Columbia's final flight was called STS-107. Seven astronauts were assigned to perform a mission that was seen as one last return to NASA's older days of monotonous microgravity research. The agency had flown successfully for years following the *Challenger* disaster in 1986. *Columbia's* doomed mission would expose NASA's return to bad habits, like ignoring warnings of possible danger from engineers. But that would come later. For the astronauts who were chosen, it was a flight, and that was enough.

Being named a member of an astronaut class is just the first challenge. After two years of basic training, candidates get in line for their first flight assignments. William McCool sat and waited as classmate after classmate was included in high-profile space missions. Astronauts Jeff Williams and Dan Tani, part of McCool's class, got to do spacewalks outside the International Space Station. When *Columbia* blasted off on STS-107 on January 16, 2003, McCool's classmate Don Pettit was in the middle of a long-term mission aboard the International Space Station. McCool would be on that shuttle crew, and it would be his name getting stitched on a crew patch for a mission in space. The mission patch for *Columbia* flight STS-107 would be in the shape of the shuttle itself, with McCool's name placed at the bottom of the right-hand wing. Designing a crew patch was one of those things astronauts did in preparation for a space mission, like posing for a group crew photograph, visiting factories where shuttle components were built, and taking questions from reporters, although rookie pilots frequently were ignored by the press.

Still, McCool was glad to go, as were the other crew members from the class of 1996, David Brown and Laurel Clark. *Columbia* commander Rick Husband had been part of the first docking mission to the International Space Station. Michael Anderson had visited Mir, and Kalpana Chawla flew aboard *Columbia* during a trouble-filled microgravity science mission in 1997. These were the veterans on the flight.

The seventh member of the crew was someone NASA initially didn't talk about by name. Ilan Ramon's inclusion on the STS-107 flight apparently began with an agreement forged by President Bill Clinton and Israeli prime minister Shimon Peres.[1] The Clinton administration was used to using space as a tool of diplomacy. The International Space Station was pushed as a way for countries around the world to work together on a big nonmilitary project. Putting an Israeli on a short shuttle mission fit in with that philosophy.[2]

The space agency admitted to the press that there would be a payload specialist on *Columbia*'s sixteen-day science flight. Weeks later there was mention that this person was from Israel. Still later they specified that his name was Ilan Ramon. "When I first came to the Johnson Space Center," said Ramon, "I remember hearing English, Russian, French, Japanese. I felt like I was truly part of an international community." Ramon was well known to the people of Israel as one of the F-16 jet fighter pilots who helped destroy the Iraqi nuclear reactor at Osirak in 1981. During his space mission, he would operate a telescope aboard *Columbia* designed to study dust storms from orbit. Ramon also would carry items into space including a drawing by a fourteen-year-old victim of the Auschwitz concentration camp, imagining what Earth would look like if seen by someone on the moon. The drawing depicted craggy mountains on the lunar surface with Earth in the background.[3]

Ilan Ramon wasn't the first person assigned to STS-107. The way the crew was selected gives an indication of how close-knit the NASA astronaut corps really is. Dr. John Clark is a flight surgeon at the Johnson Space Center in Houston. Lanky and with a look of seriousness about him, he recalls how his wife Laurel, a navy commander, became a member of the astronaut class of 1996. "She finished astronaut training in 1998 and almost immediately was assigned to the *Columbia* mission," he says. The flight would be led by veteran Rick Husband, and one of the rookie crew members would be Laurel Clark's astronaut classmate David Brown. John Clark had worked with Husband on his first space mission and had flown with Brown while they were in the U.S. Navy. "There's just a weird connectivity with this crew," says Dr. Clark. "I gave Ilan his preflight physical, and I knew Willie [McCool] from Laurel's astronaut class, so we grew together as a family."

The presence of an Israeli crew member prompted tight security in Houston and at the Kennedy Space Center. A traditional press briefing is held at KSC prior to every launchpad dress rehearsal. As Ramon and his crewmates gathered in front of the Press Site grandstand to take questions, SWAT officers armed with automatic weapons stood just a few feet away. On launch day, fighter jets and sophisticated radar kept watch around the launchpad in case terrorists tried to strike at the shuttle, which was loaded with a half million gallons of explosive hydrogen and oxygen fuel.[4]

Engineers See Something Wrong during Liftoff

A terrorist attack wasn't necessary to doom the shuttle. Falling foam insulation from the 150-foot-tall external fuel tank sealed the crew's fate barely a minute into the trip.

Once *Columbia* settled into orbit, the astronauts floated into the Spacehab modules where their eighty experiments were kept. While they searched for science data, the press searched for stories to write about. The novelty of Ilan Ramon quickly ran its course. That left reporters with little fodder but the arcane experiments that accompanied the astronauts to space. One dealt with creating micropellets of medicine to fight cancer. Another dealt with finding safer ways to contain hydrogen fuel without using hollow tanks. School students contributed an experiment involving golden orb weaver spiders from Australia. The astronauts would observe how the absence of gravity changed the way the arachnids wove their usually perfect orb-shaped webs.[5]

It was scintillating stuff for the scientific community, but it was also the kind of material that left news editors groaning. Mercifully, an air conditioner for the twin Spacehab modules broke down. The astronauts had to thread a long air vent from the crew cabin back into the experiment labs to keep things cool. That made for decent copy. Then, thank goodness, there was Super Bowl Sunday. The Tampa Bay Buccaneers were facing the Oakland Raiders for the annual football championship. During a televised press conference from orbit, Willie McCool, Michael Anderson, and David Brown obliged the media with predictions of the outcome.

"Since I'm from San Diego, I'll go for Oakland," said McCool, eliciting a scowl from Brown. "Oh, Dave says I'm in trouble."

"Well, since we have a lot of people at Kennedy Space Center who worked so hard to make our shuttle ready to fly," responded Brown, "I'm going for Tampa Bay."

That comment was ironic, since a debate was raging among those same engineers over disturbing video showing something falling from the tank and striking the front of the shuttle. They couldn't order the astronauts to use the robot arm to peer over the side of the spacecraft. The robot arm was removed, since this was just a science mission and no attempt to dock with the space station would be made.[6]

The astronauts, it seemed, would just have to take their chances during the long, fiery plunge into the earth's atmosphere. NASA managers ruled it wasn't a problem.

Video that miraculously survived the shuttle's breakup showed four of the astronauts in their heavy orange pressure suits, strapped into their flight seats. McCool was seen shuffling cards from the flight plan while Rick Husband drank from a small beverage bag. Some of the crew were not wearing their gloves. The Columbia Accident Investigation Board said it made no difference.[7]

At the Kennedy Space Center, reporters started showing up around seven a.m. Fog around the 15,000-foot runway was a concern for astronaut Kent Rominger, who flew a NASA training jet to judge the conditions *Columbia* would encounter. "We're happy with the weather at KSC," said capcom Charlie Hobaugh. "You are go for the burn." The "burn" is a two-and-a-half-minute firing of the shuttle's largest chemically fueled jet thrusters. Instead of circling the earth again, *Columbia* would begin the drop into the atmosphere.

The Routine Landing That Wasn't

The decision to proceed with touchdown surprised some of the assembled reporters. Mission Control was notoriously conservative when it came to the weather at the Kennedy Space Center. Storms can appear out of nowhere in the humidity that surrounds the concrete runway. Both the forecasters and the astronaut flying a NASA training jet have to be able to guarantee favorable weather during the hour after the shuttle fires its jets and heads for the sky around the Kennedy Space Center. Once the orbiter starts the trip down, there's no turning back, and conditions have to be good. If clouds or fog threaten, then a wave-off is ordered by Mission Control. That means at least another hour and a half as the shuttle circles the globe again for another try. That's the best-case scenario. The worst one is a complete scrub, where all opportunities have been exhausted and the crew has to try again the next day. That wasn't the case with *Columbia*. The astronauts were coming down, and the next move for the press was to board buses from the media center to the landing strip.

Normally, for journalists, one shuttle landing is much like another. The press bus is soon surrounded by trees as, once cleared by security, it moves down the road to the landing strip. With late-night landings, the only sign of human presence in this isolated area is the bright floodlights that illuminate the end of the runway the shuttle will use.

Once the press bus pulls up, journalists step off, swap stories, double-check their equipment. The parking lot is unpaved, and the press building is small and basic. During daytime landings in the summer, it's the only place with air conditioning where journalists can get a temporary break from Florida's heat and humidity while looking out the large windows facing the runway. Sometimes a reporter will set up shop there to write a story and listen to NASA mission audio, with their equipment perched on a window sill. A counter separates the public affairs officers from the press in the main press facility. A TV feed featuring the capsule communicator, or capcom, can be heard over the loudspeakers along with a narrator from NASA Public Affairs. The doors open onto a wet grassy field. Just beyond is the Shuttle Landing Facility, or SLF—NASA's term for the concrete runway. It's the second longest in the nation and was built specifically to handle an orbiter as it makes the steep drop toward touchdown at the end of its mission. A cluster of support vehicles gather alongside the landing strip.

There is speculation among the reporters on who might be available for impromptu interviews after the landing. The viewing area is divided into three patches of ground. The far left is for journalists, the center for VIPs invited to the landing by KSC director Roy Bridges, and the far right for the astronauts' families. Occasionally someone will walk to the VIP fence and take questions. But woe be to uninformed press newcomers who try to sneak past the white fencing to snag a comment or two from landing guests in the VIP section, much less the astronauts' relatives. They are quickly ushered back to the "press area." The only chance is to wait where the VIPs arrive. As the guests walk from their nicer buses to the viewing area, sharp-eyed reporters lining the path can often spot a former astronaut or NASA manager and ask for some tape. The scene resembles a rustic Hollywood premiere with actors strolling past the paparazzi.

Once the shuttle is cleared for landing, the process is "by the numbers." There's no turning back for the astronauts, so one hour after the jet thruster burn, the event is all over. On clear days, just minutes before touchdown the shuttle can be seen high in the sky as a small black arrowhead. The astronauts are in control during the final two minutes, with the commander using the control stick to make the glide without engine power. Sometimes his sidekick, the shuttle pilot, gets to take the controls for a short time. That gives the rookie on the flight a chance to get some "stick time" and experience the feel of the spacecraft.[8]

Between the press and the landing strip is a large lighted digital clock, slowly ticking toward zero. On *Columbia's* landing day the mood is upbeat, but the press area is largely silent. Comments from Mission Control can be heard over the television monitor, TV reporters practice their lines in front of their cameras, and the clock continues counting down.

What few can hear is the furious exchange going on in Mission Control between *Columbia*'s Leroy Cain and some of his colleagues, notably Jeffrey Kling. Each person on the landing team has a job, and that job carries a nickname to be used over the communication loops. Cain is flight director, so coworkers refer to him as Flight. Kling is *Columbia*'s Maintenance, Mechanical, Arm, and Crew Systems officer, or MMACS (pronounced Max) for short.

Seconds after *Columbia* lifted off, a section of orange foam insulation tore off the external fuel tank and struck the shuttle's left wing. Or, more specifically, *Columbia* hit the foam. Investigators found that the spacecraft was accelerating at the moment of impact, so the shuttle flew into the debris, breaking one of the gray heat shields on the leading edge of the wing. Video taken by the *Columbia* crew inside the flight deck that survived the accident will show them commenting on the early effects of reentry. The shuttle's black belly tiles scrape against the air during the trip down. The friction generates 3,000 degrees of heat, which forms bright colors visible from the windows of *Columbia*'s cockpit. The spectacular light show is usually harmless. This time, the superheated air is leaking through the broken heat shield, melting the aluminum structure of the shuttle's left wing and tearing it off. Sensors are being incinerated, and that creates confusion on the ground, as one by one, the indicators go dead.

"Flight, MMACS," says Jeff Kling, asking to talk with flight director Leroy Cain.

"Go ahead, MMACS," responds Cain.

"FYI, I've lost four separate temperature transducers on the left side of the vehicle," says Kling emotionlessly while looking at his computer screen. "Hydraulic return temperatures. Two of them are on system one, and one in each of systems two and three."

"Four hydraulic return temps?" asks Cain quizzically.

"To the left outboard and left inboard elevons," Kling specifies.

The elevons are the wing flaps that commander Rick Husband and pilot Willie McCool will use to bring *Columbia* down onto the runway. For now, the computers aboard the shuttle are controlling the descent. As the left wing begins to disintegrate, aerodynamic drag changes and the spacecraft begins to shift to one side. The onboard computers sense the problem and fire *Columbia*'s jet thrusters in a desperate tug-of-war to stay on course. The end is near.

"You're telling me you lost them all at exactly the same time?" asks Leroy Cain from his console in Mission Control.

"No, not exactly," says Jeff Kling. "They were within probably four or five seconds of each other."

"Okay," says Cain. "Where are those, where is that instrumentation located?"

"All four of them are located in the aft part of the left wing," says Kling. "Right in front of the elevons, elevon actuators. And there's no commonality."

Cain continues to quiz Kling on the growing sensor problems on *Columbia*'s left side, meanwhile checking on other routine elements of the landing, like the weather and the navigational beacons ready to beam information to astronaut Rick Husband during the final critical moments before touchdown. The tone of Cain's and Kling's voices becomes more urgent as the problems mount and data on Jeff Kling's screen point to something that can be deadly.

"Flight, MMACS," says Kling.

"Go!" says Cain.

"We just lost tire pressure on the left outboard and left inboard, both tires," says Kling.

What was once a series of confusing signals is now clearly a threat to the crew. Neither Mission Control nor the astronauts have been speaking as if they knew how bad the situation really was. No indicator light in Houston reads "broken heat shield," and the shuttle crew can't see what is happening. The spot on the left wing, identified as the place where the foam hit and the shield broke, is the spot where the wing bows out from *Columbia*'s hull. It's behind and below the crew cabin, so it is unlikely the crew can crane around to see what is killing them.

Blown tires on a shuttle landing would be a crisis for NASA. *Columbia* has three sets of landing gear to cushion the drop down onto the runway in Florida. One small set pop out under the vehicle's nose, and larger sets are under either wing, forming a tricycle of tires beneath the craft. From Mission Control's perspective, *Columbia*'s left set of wheels is clearly compromised.

At this point capcom Charlie Hobaugh joins in the drama by communicating with *Columbia* commander Rick Husband. Their exchange, which the public can hear, is the first obvious indication that *Columbia* is in serious danger. "And, *Columbia*, Houston," says Hobaugh. "We see your tire pressure message and we did not copy your last [message]."

"Roger, uh—" responds Husband until his voice is cut off.

There are bursts of static on the line, punctuated by moments of silence. "*Columbia*, Houston, comm check," asks Hobaugh. The capcom continues to ask *Columbia*'s crew to respond. His deadpan messages are met with silence, or static known in NASA jargon as "ratty comm." Cain soon orders his crew to lock the doors, telephone no one, and conserve all of their data.

The *Columbia* disaster has occurred.[9]

The loss of communication between Hobaugh and the astronauts prompts confusion among NASA workers and the *Columbia* families at the Kennedy Space Center's landing strip. Speculation begins immediately on whether the spacecraft is missing or is perhaps about to land elsewhere and require a rescue operation.

The problem had become truly apparent to those watching the digital clock. The loss of communication between Houston and *Columbia* occurred sixteen minutes before the planned landing. At that point it seemed that *Columbia*'s radio might be out. But now, two minutes before touchdown, the silence is deafening.

Two minutes before a shuttle lands, there are two sonic booms, sounding like a double shotgun blast. The shock waves come off the nose and tail of the vehicle, and they signal the beginning of the touchdown. Moments before *Columbia*'s scheduled touchdown, the lighted digital countdown clock between the press and the runway has ticked down to the two-minute mark and then kept on going. No sonic booms. One NASA manager states matter-of-factly, "We've lost an orbiter."

The Press Catches Wind of the Columbia Disaster

The situation slowly became clear across the country, including in the studios of National Public Radio in Washington, D.C. Listeners to *Weekend Edition Saturday* were hearing a commentary from Daniel Schorr, a former colleague of broadcasting great Edward R. Murrow. That segment was pretaped, and as it played, senior editor Gwen Thompkins had a chance to breathe and thank her lucky stars that the lead story for the show had worked out. Her main guest hadn't been where he was supposed to be. "The guy was in a grocery store!" she says with consternation.

Thompkins is referring to Philip Stephens, a columnist for the *Financial Times* of London. The morning of *Columbia*'s landing, the United States and Great Britain were on the verge of war with Iraq. Prime Minister Tony Blair was about to ask the United Nations for yet another resolution against Saddam Hussein. Stephens was supposed to be interviewed about it by *Weekend Edition* host Scott Simon. "I finally caught up with him on his cell," says Thompkins. "He was in a London grocery store—I think he was with his kid—and he did a great job for us. I went away feeling this was going to be great."

Saturday mornings at National Public Radio are usually quieter than that. A skeleton crew works on the weekend shows, and the crowds of people that fill the hallways from Monday to Friday are gone. Thompkins and senior producer Peter Breslow had the place pretty much to themselves. With Philip Stephens back to his grocery shopping in London and Daniel Schorr's commentary playing as scheduled, Thompkins and Breslow returned to the maze of cubicles and desks in the NPR newsroom that they shared with the staff of the weekday *Morning Edition* program. Behind the desks were tiny studios where a small number of editors were putting together four-to-five-minute radio stories by mixing audio tape of reporters reading their scripts, comments from people they'd interviewed, and background sound to give their stories a sense of place. Thompkins and Breslow settled down at their computer screens to review wire service stories and bulletins. About 9:15 a.m., Thompkins saw something strange.[10]

"*Columbia* late," the wires said.

"For us, it wasn't a big deal," recalls Thompkins. "There had been so many landings of these shuttles, and there was no doubt that *Columbia* would land and the world would go on." That first bulletin didn't put Thompkins and Breslow into the mindset that something serious had happened to the shuttle and the astronauts. Then the second and third bulletins blipped by on their computer screens. "Clearly the wires were trying to tell us something," says Thompkins. "But they were going with only what they had." Then the television overhead began to chime in. The

Weekend Edition crew left that set tuned to CNN. The dribs and drabs of information began forming a picture of a tragedy, and then a wire bulletin made the situation plain.

"*Columbia* lost," it said.

Thus began the process that editors fear and relish the most. Breaking news was happening, and Thompkins set into motion unscripted coverage that would last through the morning and into the evening. Informing host Scott Simon of the tragedy was job number one; filling the next three hours until the afternoon crew took over was next.

Producers on NPR programs always think five minutes ahead. While the on-air anchors are conducting interviews and introducing prerecorded radio reports, people like Gwen Thompkins and Peter Breslow are sweating what to put on after that. *Columbia* was gone, and Scott Simon was getting information from the Kennedy Space Center, but that could go on only so long.

NPR's next bit of information came from an unlikely location. It was a Washington, D.C., beauty parlor where Darcy Bacon was having a pedicure.

Bacon is the booker for *Weekend Edition Saturday*, although she prefers the title "associate producer." That means Tuesday through Thursday she helps find stories for the program and tracks down people to be interviewed. By Saturday her work on *Weekend Edition* is over, though she usually tunes in like everyone else to hear how the various segments turned out. The morning *Columbia* was about to land, as she settled in for her pedicure, she asked the stylist to turn the radio to NPR so she could listen. "I was there, sitting in the chair," says Bacon. "And I heard Scott [Simon] give this sober, terse announcement that they had lost touch with the space shuttle." Even with her feet soaking, Bacon kicked into booker mode. She knew Breslow and Thompkins would be looking for guests to fill the airtime, so the Rolodex in her head started flipping and she remembered a visit to a friend's house up in Maine.

A neighbor of that friend was former astronaut Rick Hauck, who had led the crew of space shuttle *Discovery* on NASA's first mission following the *Challenger* disaster of 1986. "I knew him from there," says Bacon. "I just knew him as someone who was thoughtful, and the space shuttle was obviously a big part of his life." With the stylist looking on, Bacon was on her cell phone tipping off the NPR producers where to find Hauck. Within three minutes, Hauck's voice was pouring out of the radio. Being a part of the story was gratifying, but Bacon still felt the confusion and pain as the disaster unfolded. "All we knew initially was that they'd lost contact with the shuttle," she says. "For those of us old enough to remember *Challenger*, you knew that any abrupt change like that had serious implications. So you held your breath and hoped for the best, but not really expecting it. You didn't have any idea how it would turn out, until about fifteen minutes later when they started finding pieces [of wreckage]."[11]

Thompkins considered Bacon's call a lucky break. The Hauck interview was adding good content to the ongoing radio coverage. Reports on television were now referring to burning debris streaking across the sky of East Texas. While Bacon's contribution was something NPR expected, what happened next was not. "Listeners started calling," says Thompkins. That didn't happen during *Weekend Edition*. NPR had call-in programs where the host would repeat the number a few times to prompt people to phone and offer questions and comments. No number was given during the coverage of the *Columbia* disaster. Residents of East Texas called anyway. "These were people who lived along the trajectory of the descent, the path where *Columbia* was coming down," says Thompkins. "They'd say they heard a terrible noise, or things were falling as the shuttle reentered the atmosphere. It was a spontaneous thing, where people felt they were as much a part of the story as we were. That was way cool!"

Soon afterward, NASA's recovery effort began. Little hope existed that any of the astronauts could have survived, since *Columbia* was traveling at 10,000 miles per hour at the height of the fiery reentry when the spacecraft broke up. The concern turned to the public, especially around towns like Hemphill, Texas. Everyone was warned to stay away from debris on the ground. Sharp, jagged edges on the wreckage were a major concern, as were the explosive chemicals possibly mixed in with the debris. One worry that emerged at the time of the *Challenger* explosion in 1986 was the possibility that, even with the shuttle's wreckage at the bottom of the Atlantic, the fuel tanks from its jet thrusters could wash ashore and someone might get too close. Each tank was filled with one of two self-igniting chemicals. That same threat could be present following the *Columbia* accident in 2003. Nitrogen tetroxide and monomethyl hydrazine were each very poisonous and corrosive. And if the two chemicals from the separate tanks leaked out and made contact, they would explode.

NASA Begins the Recovery Effort

Some members of NASA's astronaut corps back in Houston helped co-ordinate the search for wreckage and for the remains of their fallen colleagues. Others had the job of consoling family members of the *Columbia* crew. Still others would field questions from reporters covering the tragedy. Two astronauts who would find themselves in key roles after the accident were Eileen Collins and Steve Lindsey. These space veterans were in line to command the first and second shuttle flights following *Columbia*. Now they and the fleet were grounded while NASA worked to recover from *Columbia*.

Like everyone at NASA, Lindsey and Collins were dealing with the pain of losing a shuttle and its crew. Collins was at home in Houston when word came. The astronaut supervised her two-year-old son in the kitchen as he occupied himself playing in the sink while she fielded calls from NASA and her crew. Lindsey was in the middle of the crisis. "I was actually at the landing site watching," he recalls somberly. "I knew something was wrong—loss of communication, which never happens—I knew something was wrong. I didn't know what, but I knew." The difficulty didn't stop there. Lindsey was assigned as a family escort and had to take the tough questions as the tragedy came into focus for NASA as well as for the nation. The country had lost a group of seven people who smiled out from their official crew portrait. Lindsey had lost six colleagues and a crewmate.[12]

Steve Lindsey Remembers a Lost Crewmate

Kalpana Chawla and Lindsey shared the glory of their first rookie blastoff and took their share of the heat when things went bad.

Both were aboard *Columbia* in 1997 for mission STS-87. One cargo was a box-shaped satellite called Spartan-204. Chawla was the robotic arm operator with a two-part job. First she would lift Spartan out of the payload bay, switch it on by remote control, and release it. The shiny gold package was supposed to gather data and then be scooped up by the shuttle so scientists could download the information back on Earth.

Something went wrong.

After its release, Spartan was programmed to do a pirouette to signal that it was working right. Spartan just hung there in space. Chawla was

Figure 40. Crew portrait of shuttle mission STS-87. Front, left to right: astronauts Steve Lindsey and Kevin Kregel. Back, left to right: astronauts Kalpana Chawla (who later died in the *Columbia* accident) and Winston Scott, Japanese astronaut Takao Doi, and Ukrainian cosmonaut Leonid Kadenyuk. Courtesy of NASA.

trying to grapple it with the arm. She bumped it, sending the satellite into a tumble. Spacewalkers finally retrieved Spartan, but the stigma of a failed mission remained. The STS-107 mission aboard *Columbia* was Chawla's chance to put that earlier flight behind her. She didn't live to see the outcome. "We talked to the 107 crew every day," said Lindsey. "And thirty percent of the science made it back after the accident. She did a fantastic job on the flight. They did eighty experiments, she did have a great mission."

After his trouble-filled rookie flight, Lindsey and his next shuttle crew made up for the Spartan problem by carrying the same satellite into space the next year. Few in the press took note of the quiet Spartan reflight. Most of the attention focused on one astronaut who flew on the mission, Mercury pioneer John Glenn. "One of my most memorable moments was watching him look out the windows for the first time after thirty-six years," says Lindsey. "And seeing the earth after that many years."

Monday Morning at the Kennedy Space Center

Since *Columbia* was lost on a Saturday, for many Kennedy Space Center workers the first time to commiserate with friends and colleagues came on Monday. That included office manager Sheryl Chaffee-Marshall, who worked in the KSC headquarters office. When space shuttle *Challenger* exploded in 1986, friends and colleagues had dropped by, protectively, to make sure she was okay. "I saw it [*Challenger*] happen," says Chaffee, "and it was hard." Many people witnessed *Challenger*, and many others saw video images of *Columbia*'s fiery wreckage streaming through the sky. But for Chaffee the aftermath of both of NASA's shuttle disasters was different. One clue to why is the color photograph hanging on her office wall. Carefully framed along with a cloth mission patch from *Apollo 1*, the picture shows three men in heavy white spacesuits standing in front of a launch gantry. The man on the right is astronaut Roger Chaffee, Sheryl's father, who perished in the Apollo launchpad fire in 1967. People could talk about what it was like to lose an astronaut in the line of duty. Sheryl knew firsthand. Around her neck, on a chain, is a small gold heart. "I always wear it," she says. "My mother gave it to me. My father was going to fly it on his mission."

Navy lieutenant commander Roger Chaffee was thirty-one years old when he was assigned to fly the first test mission for the Apollo command capsule that would take men to the moon. It was to be his rookie flight in orbit, and he would have two crewmates. One was astronaut Ed White, who flew the second mission of the two-person Gemini capsules in 1965. White made America's first spacewalk, floating briefly in a silver spacesuit outside the protection of his spacecraft. The crew was rounded out by a man many considered a space legend. Virgil "Gus" Grissom, one of the original seven Mercury astronauts and commander of the ill-fated flight of the Liberty Bell 7 capsule that sank after splashdown in 1962. Grissom put that mishap behind him in 1965 with the successful inaugural flight of the Gemini capsule nicknamed Molly Brown, after the woman who survived the sinking of the Titanic.

Training and other NASA business kept Chaffee and his crewmates on the road much of the time. But Christmas in 1966, before the fateful launchpad test, was different. Chaffee was home in Houston a lot more, and his daughter remembers him being handy around the house, especially decorating for the holidays. "This was the 1960s, and you couldn't just go to Home Depot and buy reindeer to go in the yard," says Sheryl Chaffee. "So my dad built things that flashed." The neighborhood garden club held a contest for the best Christmas display, and the Chaffees' won that year.

For the launchpad test on January 27, 1967, Grissom and Chaffee and White had to be suited up as they would be on liftoff day. The astronauts were sealed inside the crew cabin with its pure, and flammable, oxygen atmosphere.

Sheryl Chaffee was eight years old when she and her brother were called into the bedroom by their mother. This family gathering had been preceded by a short private visit by future *Apollo 11* astronaut Michael Collins. Sheryl's mother explained that their father wasn't coming home anymore. "I thought they were getting divorced," says Sheryl. "I was eight years old, I didn't even know how I knew what divorced was." In fact, on the launchpad in Florida, fire had erupted inside the Apollo capsule. NASA engineers rushed to the gantry, but it was too late. The three astronauts were gone. Sheryl Chaffee remembers the funeral, and seeing her father's casket. "I had nightmares," she says. "I saw him coming home, and he was covered in Band-Aids."

Years later, in 1983, Sheryl and her husband moved to Brevard County, Florida. She needed a job and applied to the Kennedy Space Center for an office position. It was far from the shuttles, so there was a comfortable emotional distance from the kind of occupation her father pursued to the end. The *Challenger* accident, three years later, changed that. And now more families were in mourning as the fate of the seven *Columbia* astronauts became clear on February 1, 2003.

"I tried to do whatever I could," says Chaffee. "To get the word out to the families, that you can go on. It's hard, but you can go on."[13]

For NASA, after the *Columbia* accident, it would take more than consoling, as the agency braced itself for a renewed flood of questions and criticism surrounding its operation of the shuttle program.

NASA Prepares to Answer Difficult Questions

Engineers from the Kennedy Space Center in Florida and astronauts from the Johnson Space Center in Houston boarded cargo planes to head to the crash site. The search area was too big for NASA to handle by itself. Volunteers were needed, a lot of them.

While this recovery effort began, a new panel had been formed to investigate the *Columbia* disaster, and NASA would have more embarrassing questions to answer. The agency had been through a previous tragedy that called management procedures into question along with shuttle hardware. NASA said they had fixed the problems. But apparently they had not fixed them all.

11

The Aftermath

Before the *Columbia* disaster, the space shuttle had become indispensable in the effort to build the International Space Station. Interest in the program revived, after years of unexciting launches on routine-seeming errands, as NASA struggled to find a specific mission to give the shuttle focus. Now its indispensability proved to have a double edge. All of the spacecraft were grounded following this second disaster, and that left the space station in a serious bind. The United States was without a way to fly to orbit, which meant construction on ISS would come to a halt.

NASA quickly asked the Russian Space Agency to keep astronauts and cosmonauts flying to and from the outpost on its Soyuz space capsules. Some resupply was possible with the robotic vehicles called Progress ships. The decision, back in the mid-1990s, to put the proposed new station on an orbit over both Florida and Russia's launchpads turned out to be fortuitous, since that put ISS within Moscow's reach. But without the shuttle available, there would be no new compartments or solar panels on the complex. Also, the smaller Russian spacecraft couldn't return trash and used equipment to Earth, so things started getting cluttered as station crews crammed expended gear wherever they could.

While the crew members aboard ISS worked out their own problems, NASA had concerns as well. At one time, federal managers would line up for advice from NASA. The agency was known in the 1960s for doing the impossible with incredible organization and well-publicized results. Following *Columbia*, Washington decided NASA was, once again, the one that needed help. The agency that had staged the Apollo moon landings would now sit and listen as others spelled out in excruciating detail how NASA had fatally erred.

The Columbia Accident Investigation Board

After the *Challenger* accident in 1986, when the Rogers Commission was assembled to conduct an investigation, a member of that panel was Dr. Sally Ride, the first American woman to fly in space. Now, in 2003, she would be a member of the Columbia Accident Investigation Board, also known as the CAIB. Those joining Ride in the inquiry were less well known but considerably qualified. Admiral Harold Gehman, who would lead the board, had served as NATO's Supreme Allied Commander, Atlantic, and had cochaired the investigation into the terrorist attack on USS *Cole*. Dr. Sheila Widnall was a former secretary of the air force and the first woman to head a department at MIT. Scott Hubbard directed NASA's Ames Research Center and had led the investigation into the two spectacular failures of the agency's Mars Polar Lander and Mars Climate Orbiter. During the *Challenger* probe, investigators found management problems as well as mechanical issues at NASA, and the CAIB was expected to use a similar brush to tar and feather the U.S. space program once again.[1]

The thirteen members of the CAIB would break up into subcommittees to study the data from the *Columbia* accident and, as a group, would listen to witnesses from NASA, its contractors, and academia as they searched for answers. There would be questions over reports by engineers that launch-day video clearly showed debris coming off the external fuel tank and striking the left wing, leaving a puff of dust. How far up the chain of command those reports went, and how mission managers responded, were key concerns of the CAIB. Also, there had been early suggestions that military spy satellites should turn their cameras on *Columbia* while it was still safely in orbit to scan for damage. What happened with that idea? Following the accident, ground-based radar images of the shuttle were reviewed and revealed a small mysterious object floating away from the doomed spacecraft two days after launch. Was it a piece of loose equipment from the open cargo bay, or a piece of debris left in orbit by an earlier spaceflight that somehow impacted *Columbia*?

These were all new problems directly associated with the breakup of the shuttle. But a disturbing old problem was becoming clear as well. In 1986, during the investigation into the loss of *Challenger*, NASA was criticized for not acting more decisively when problems began appearing in the critical field joints of the solid rocket boosters. The agency had examined evidence that hot exhaust gases could eat away at the asbestos putty that lined the seam between adjoining rocket casing sections, but there were no definitive solutions. The rubbery O-ring gaskets that were supposed to seal these joints were found to be damaged following the second launch of *Columbia* and a classified mission featuring shuttle *Discovery* in 1985. Instead of grounding the fleet and resolving the problem, NASA ordered modifications but kept on flying. Soon after that, the *Challenger* accident occurred. Now, in Houston in 2003, the board investigating the *Columbia* disaster heard testimony that a similar mindset persisted.[2]

Dr. Harry McDonald was asked to speak before the CAIB on March 6 to address how NASA reacted to a previous near disaster during a shuttle mission. McDonald was former director of NASA's Ames Research Center, and the panel wanted to hear his experience investigating the trouble-filled liftoff of *Columbia* in 1999, four years before the craft was lost during reentry. During this probe, McDonald crossed paths with someone who would be at the heart of the effort to resume shuttle flights after *Columbia*. That was astronaut Eileen Collins, who was in line to command the next mission following the accident. Collins had been the first female shuttle pilot and, in 1999, became the first woman to command a mission. Her five-member crew narrowly avoided becoming the subject of a disaster investigation of its own, but had "dodged the bullet." McDonald looked into what happened on that earlier flight, and why those lessons didn't save the astronauts of *Columbia*'s last mission in 2003.

After being selected as an astronaut candidate in 1990 and doing her two years of basic training. Collins was assigned to pilot *Discovery* in 1994 on the first rendezvous mission between the shuttle and the Russian space

Figure 41. Astronaut Eileen Collins following her flight as NASA's first female shuttle commander in 1999. *Columbia* lost two main engine controllers and suffered a hydrogen leak, which prompted a premature main engine shutdown. *Columbia* limped into an orbit seven miles lower than what NASA wanted. Courtesy of NASA.

station Mir. She piloted *Atlantis* on a Mir docking flight in 1997, and then took command of *Columbia* in 1999 with the Chandra X-Ray Observatory as its main cargo. A few seconds after liftoff in the summer of 1999, a warning indication flashed on Collins's dashboard. *Columbia* had experienced a short circuit, losing two main engine computers, and was leaking hydrogen fuel.

Onlookers on the ground were tipped off by the heightened amount of radio chatter between the shuttle and Houston that something was wrong. Collins and Mission Control kept making matter-of-fact exchanges by radio about what wasn't working on the shuttle and what was. There was a premature main engine shutdown, but the astronauts managed to limp into an orbit seven miles lower than NASA wanted. They had to use backup thrusters to do it. Still, they deployed the Chandra observatory and landed back on Earth safely.[3]

Columbia was found to have a number of problems. Frayed wiring inside the spacecraft was blamed for shorting out the two main engine computers. What appeared to disturb the Columbia Accident Investigation Board more was the mechanical failure related to the hydrogen fuel leak, and what NASA did about it. Video during *Columbia*'s 1999 launch showed a stream of explosive hydrogen fuel squirting from one engine. It turned out that a pin inside the motor's fuel injector had broken loose. The supercold liquid propellant that fuels the shuttle's motors also helps cool the system. The liquid hydrogen runs through tubes wrapped around the black bell-shaped nozzles, which belch exhaust during flight. When the pin came free, it traveled through the cooling tube and weakened it. That led to a rupture, and the resulting loss of coolant allowed the engine to overheat and shut down prematurely.[4]

Harry McDonald was asked to look into *Columbia*'s 1999 problems as part of a group called the shuttle Independent Assessment Team, or SAIT. During his testimony before the Columbia Accident Investigation Board in 2003, McDonald spoke about the mixed messages he received from NASA management on how the problem should be handled. He told the CAIB that NASA's top administrator, Dan Goldin, urged him to leave "no stone unturned" to learn the cause of the anomaly. The shuttles were grounded and the miles of wiring checked and repaired before launches were resumed, but McDonald wasn't satisfied with the resolution of the engine pin problem. The SAIT wanted engineers to track down the exact cause of the loose pins before any more flights occurred. NASA disagreed. In fact, the agency went before Congress in September of 1999 and told the House Subcommittee on Space and Aeronautics that the pin situation did not represent a danger. "I was personally disappointed that NASA didn't or wouldn't follow up on our recommendations," McDonald said.

There were nineteen previous instances of engine pins coming loose on blastoff prior to *Columbia*'s 1999 launch, and McDonald believed that NASA thought the track record of successful flights made the situation less dangerous. CAIB member Major General Kenneth Hess, the air force's chief of safety, asked McDonald to comment on NASA's apparent tendency to accept risk in the shuttle program and how that mentality may have played into the *Columbia* accident. McDonald responded that there was a basic flaw in the thinking among NASA managers and engineers. If a specific threat is judged to pose a one-in-a-hundred chance of causing a shuttle disaster, McDonald argued, the possibility of a disaster is equally serious on every flight, from the first one to the last. But NASA, he said, felt that the threat lessened after each successful shuttle mission because nothing had happened up to that point. "There seemed to be the perception within the agency that if I have flown twenty times, the risk is less than if I have just flown once," said McDonald. "And we were continually attempting to inform them unless they've changed the risk positively, you still have the same issue even after fifty or sixty flights." McDonald believed that NASA was safety conscious, but that there were holes in the system, particularly when it came to waivers signed on launch day to keep a countdown moving toward liftoff. He told the CAIB that launch managers would often sign waivers to bypass technical concerns, not on the basis of firsthand knowledge, but on the recommendations of others. A new system of inspection was required, McDonald felt, not simply hiring more inspectors. Members of the CAIB then sought to see how NASA's management practices during the shuttle flight Harry McDonald examined in 1999 could have affected the way the agency handled *Columbia*'s fatal flight in 2003.[5]

On March 25 at Cape Canaveral, Florida, the board convened to hear witnesses including retired general Aloysius Casey. The thirty-four-year veteran of the air force helped develop missiles like Minuteman 2, Minuteman 3, and MX, better known as the Peacekeeper. He finished his military career as commander of the air force's Space Division, then became an aerospace consultant. General Casey said NASA clearly violated safety margins by launching the shuttle after tank foam was found to have torn off on missions prior to the *Columbia* disaster. The agency simply didn't know enough about the problem. "There is no way you can say you're operating within safety margins," Casey contended, "if you have an unknown mass impacting the aerodynamic surfaces [of the shuttle] and it has unknown damage." He compared the problem of foam falling from the external fuel tank to the leaking booster rockets that led to the *Challenger* disaster in 1986. In both instances, Casey believed, NASA kept flying shuttles despite potential problems that the system wasn't designed to withstand. In the case of *Challenger*, NASA launched even though it was very cold the morning of liftoff and the boosters hadn't been tested to see what effect freezing weather would have. Casey said NASA did tests to see what damage might be done by falling chunks of foam insulation, but not under the severity of the foam that *Columbia* faced on its fatal mission. The shuttle, in Casey's view, wasn't designed to withstand an impact like the one *Columbia* sustained, and yet no modifications were made. As to a remedy, Casey suggested finding a way to keep foam from falling in the future, such as reducing the amount of foam applied to the tanks. He cautioned, however, that trying to anticipate every fatal scenario would be difficult. "It is impossible, in my mind, for a system as large and as complex as is the shuttle, to identify with any certainty the next probable failure mode," said Casey. "So, if you go around just trying to redesign the subsystems from today's baseline, you may well spend a lot of money on things that are not really the next most probable cause."[6]

Along with expert testimony, the inquiry into the *Columbia* disaster sought clues from the debris field in East Texas where wreckage from the shuttle rained down after breakup. One critical part was a data recorder aboard *Columbia* that NASA hoped had survived the crash. If so, the magnetic tape machine could shed light on what was happening to the spacecraft before, during, and possibly after it broke up. The big break in this part of the investigation would come through the efforts of a Central Florida forest ranger.

NASA Finds Art's Part

In 2003 Art Baker worked for the State of Florida in rural Lake County, west of Orlando. Like many long-term residents, Baker was a kid who grew up in the shadow of Project Apollo. News of the *Columbia* disaster was still ringing in the ears of the nation when Baker sat down for a trim at the local barber shop. "There was a young man sitting next to me saying how he was going to the search area to help out," says Baker. "I told my boss about it, and he said did I want to go, and I said yeah."[7]

The demise of the manned moon program in 1972 occurred as Walt Disney World opened its gates and changed the Central Florida landscape forever. But the communities that surround Orlando, like Tavares, Eustis, and Umatilla, still retain their small-town charm. News that a Carrabba's restaurant is going to open is a big deal here. Billboards along Interstate 4 colored in black, orange, and yellow advertise Florida Citrus Center, where juice and fruit can be bought and shipped. The signs give a 1950s flavor to the area.

Art Baker's job as a forest ranger was watching for wildfires and driving heavy trucks. His workday often included climbing the rickety steps of a fourteen-story-tall fire watchtower and then squeezing his burly form, in a green and tan uniform topped by a green baseball-style cap, through a small wooden hatch into an un-air-conditioned observation room that afforded a view of the tree-covered carpet below. Baker still recalled how the windows in his boyhood home rattled as the big Saturn rockets carried men like Neil Armstrong and Buzz Aldrin to the moon. To many Central Floridians, NASA was family, even if they didn't work there.

One hundred thirty Florida rangers volunteered in March of 2003 to help with the shuttle recovery operation, and Art Baker was one of them. He and his coworkers from the Florida Division of Forestry boarded trucks and headed to eastern Texas where most of *Columbia*'s wreckage was scattered. The convoy arrived late that night. Search teams wouldn't be put up in local hotels. Instead they pitched their tents near the search zones where NASA wanted them to look. "They kept the lights on for us, until we settled in," says Baker. The next morning, NASA leaders staged a briefing on what the volunteers should look for. The hilly terrain in the search zone would make the job tougher. There was also a light dusting of snow, which might cover up vital clues.[8]

A small room at the local VFW was used to display debris that had already been found—circuit boards, bits of fireproof cloth, switch panels, twisted pieces of metal. "You were excited about seeing it," says Baker. "Until you left, and then you felt bad because you remembered the tragedy." The process and location of the search effort were both familiar to Baker. The oak trees here were similar to the ones that grew around Lake Griffin and Lake Eustis near his home in Florida. He had also been out to help with wildfires in California on two previous occasions. NASA used the same military-style briefing and search team patterns, where men would line up about twenty feet apart and walk as a unit along a search grid. "It was sort of hard to mess up," says Baker, "because you were pretty much plugged into what they wanted you to find."[9]

NASA saw *Columbia*'s breakup like everyone else, in the form of flaming wreckage plunging to the ground on TV. To unravel what happened, the agency needed clues. Bits and pieces recovered from the debris field would become parts of the puzzle. Telemetry data radioed to the ground during *Columbia*'s fatal plunge would reveal more about the temperatures, pressures, and stresses the space plane endured during its final moments. But the Holy Grail was something called the orbiter experiments recorder.

The OEX recorder was a thirty-five-pound box built in the 1970s, and *Columbia* was the only shuttle to have this system. The box's nine thousand feet of reel-to-reel tape would record sensor data but not transmit it anywhere. The astronauts were instructed to switch on the device a half hour before the vehicle broke up. Investigators were also certain the recorder kept running as *Columbia* disintegrated, and even a few seconds of extra readings could make a huge difference. Search team members like Art Baker were told to look for something resembling a VCR. It was slow going. "We found a big section of switch panel one day," recalls Baker. "NASA said that was important because the position the switches were in could tell a lot about the last thing the astronauts did." Baker heard stories of what other members of the Florida search team recovered, like a glove and a helmet, though he never saw those items for himself. Still there was no sign of the orbiter recorder, and the searchers knew NASA needed it. "It was kind of anticlimactic," says Baker. "You were geared up to be motivated, but you didn't talk about it much with your crewmates, because of the outcome [of the space mission]—it was sad."[10]

The snow began to clear, which prompted search team leaders to ask Baker and the other rangers to retrace their steps again and again over territory that had already yielded very little wreckage. The third time around, the group was moving up a hill and there was a log in the way. Baker climbed over it and, before looking down the way he was trained, he looked up and saw something. "It looked like a VCR," he says. "It looked like a big VCR, just the way NASA described it." Then Baker remembered his instructions upon finding wreckage. Run up to it, but don't touch it. Just call it out and wait for NASA. "I remember yelling, 'Part! Part!'" says Baker. "Then one of the guys came up and wrote my name on it with a stick. He said, this is Art's part. I said, well, okay!" It looked like the box NASA wanted, the orbiter experiments recorder, but there was no way to tell. It was battered and gray from the fiery breakup of the shuttle. Rather than wait around, Baker and his team had to move on to look for more wreckage. NASA managers arrived and took over the site. Later on, when his group was taking a break, Baker saw someone from the space agency walking by with the box wrapped in clear plastic. He knew it had to be something important. The spot where they found it was three miles from the main road, and it looked heavy. Nobody was going to lug something like that for such a long distance unless he had to.[11]

Studying the Debris Back in Florida

The recorder, along with everything else found by the search teams, made its way by flatbed truck to the Kennedy Space Center. A hangar used for unmanned satellites had been turned into a high-tech mortuary for *Columbia*'s remains. The Columbia Accident Investigation Board needed to know what all this wreckage revealed about the shuttle disaster, but NASA engineers would need to inspect and analyze the debris first. The OEX recorder that Art Baker found was the centerpiece for the poignant display of what was left of NASA's oldest orbiter.[12]

When *Challenger*'s wreckage was recovered in 1986, it looked like parts of a shuttle. The words *United States* were perfectly legible on a section of the spacecraft's main hull. Following the *Columbia* disaster in 2003, investigators had to piece together shattered and twisted debris before anything looked like the orbiter. The gray heat shields from the left wing were mounted in a plastic frame to re-create the wing's leading edge. The day reporters were taken through the hangar was the same day investigators declared that the breach that led to *Columbia*'s violent end had occurred at shield number 8. A hole punched into that shield was clearly visible through the transparent plastic of the display frame. The Columbia Accident Investigation Board would confirm these findings three weeks later.

Perhaps the most recognizable piece of wreckage was the frame of the shuttle's windshields, minus its double-paned glass. Astronauts Rick Husband and Willie McCool sat just behind these windows. NASA engineers who often gave emotionless yes-or-no answers to reporters' questions were patently moved by the sight inside the hangar. Of *Columbia*'s final moments, NASA launch director Mike Leinbach said: "She clearly put up one hell of a fight."[13]

The tape inside the orbiter experiments recorder was stretched and twisted. Still, it eventually gave investigators clues from four hundred sensors indicating that the superhot gases of reentry leaked into *Columbia*'s inner structure earlier than they first thought.

Figure 42. The "smoking gun" in the *Columbia* accident. NASA investigators loaded the reinforced carbon-carbon (RCC) heat shields from the leading edge of the left wing. This photo was taken within moments of investigators declaring shield number 8 as the spot damaged by falling foam, allowing superhot gases from reentry to blow past the breach, which led to the *Columbia* disaster. Photo by Pat Duggins.

Above: Figure 43. The windshield frame from the flight deck of space shuttle *Columbia*. Astronauts Rick Husband and William McCool sat directly behind these double-paned windows. Crew members Kalpana Chawla and Laurel Clark sat behind Husband and McCool on the flight deck. Astronauts Michael Anderson, David Brown, and Israeli Ilan Ramon were strapped into flight seats on the middeck below. Photo by Pat Duggins.

Left: Figure 44. Section of the left rear thruster pod of *Columbia*'s orbital maneuvering system (see figure 13). These black heat tiles shield the chemically fueled thrusters on either side of the tail rudder. Photo by Pat Duggins.

Figure 45. Wreckage from the *Columbia* disaster: a cluster of reaction control system (RCS) thrusters. These jets were located beneath the main windshields of the spacecraft where astronauts Rick Husband and William McCool looked out while piloting the shuttle. Photo by Pat Duggins.

All of the information gleaned from the wreckage, including the "smoking gun" from heat shield number 8 from the left-hand wing, was delivered to the CAIB. By then the panel was already amassing data on what led to the loss of *Columbia*. Its final report was issued on August 26.

The document zeroed in on the technical issues leading to the disaster, with board member Scott Hubbard summing it up this way: "In four simple words, the foam did it." The report stated that foam insulation, falling off *Columbia*'s external fuel tank, was the cause of the heat shield damage that led to the shuttle's destruction. The mysterious object that was observed floating away from the shuttle by ground-based radar installations on day two was never positively identified, although air force analysis showed it likely was a piece of the heat shield damaged during liftoff.[14]

Along with the thorough technical analysis of the accident, it soon became clear that NASA itself wouldn't get off easily.[15] Just as the Rogers Commission had stinging criticism for the role of NASA management in the *Challenger* disaster, likewise the final report of the Columbia Accident Investigation Board singled out the culture at NASA as a prominent factor in the loss of that shuttle and its crew—as prominent a factor as the foam insulation that punched a hole in the critical heat shield on the left wing. "NASA had conflicting goals of cost, schedule, and safety," said CAIB member General John Barry during a press briefing to unveil the report, "and safety lost out." Admiral Harold Gehman was meticulous in tracking down the specifics. "You find the widget that led to the accident," he said, "and the person in the fault chain that led to the failure of the widget. Then you replace the widget, and retrain or fire the person in the chain." But when it came to the *Columbia* accident, Gehman felt more was needed. Problems with NASA's culture were apparent. The desire to meet a launch schedule, possibly to the exclusion of all else, was a prime example.[16]

"We get it," said NASA administrator Sean O'Keefe following delivery of the CAIB report. The agency pledged to follow the fifteen safety recommendations from the panel. But when *Discovery* rolled to the pad for the first post-*Columbia* mission, it fell short by three. First, the CAIB wanted proven methods to repair heat tile damage before *Discovery*'s launch. NASA decided that wouldn't be possible. Definitive proof would require the astronauts to apply repair compounds to broken tiles in orbit and then return the samples to Earth so they could be put into a blast furnace. Only then could NASA be sure the repairs would work if a shuttle were really damaged on the way to space. The next recommendation was to harden the outside of the shuttle to protect it from falling debris. Not practical, said NASA, since the fleet is supposed to be retired in 2010. The third unmet suggestion was to eliminate falling debris from the external tank. NASA says it followed that recommendation where it counted. Foam insulation was removed from the top of the tank where the shuttle's nose is bolted. That's where foam tore off and hit *Columbia* in 2003.[17]

Among those who were provided with copies of the CAIB final report were the family members of the lost *Columbia* astronauts and the crew members who would fly the first post-*Columbia* shuttle mission.

NASA Remembers the Lost Astronauts

On October 27, about one hundred people sat on plastic folding chairs in front of what is known as the Space Mirror at the Kennedy Space Center. The polished black surface is forty-odd feet high and fifty feet wide. Several of the five-hundred-pound stone panels that comprise it are pierced through with names like Gus Grissom, Elliott See, and Christa McAuliffe. Each is the name of an astronaut who died while on active duty. Some of the names are familiar, like Ed White, who died in the *Apollo 1* fire with Grissom, or Dick Scobee, who was lost with McAuliffe aboard *Challenger*. Today the names of the *Columbia* crew would be officially added to the memorial.

The Astronaut High School marching band from the nearby city of Titusville was seated in a section of its own, back and to the left of the gathering. In matching white dress shirts and blue trousers, the kids played patriotic tunes for the occasion including slightly off-key renditions of "America the Beautiful" and the Lee Greenwood anthem "God Bless the

Figure 46. The *Apollo 1* crew posing for the traditional portrait. From left to right: Ed White, Gus Grissom, and Roger Chaffee. The three men died when fire broke out in the pure oxygen atmosphere of their Apollo capsule during a routine pad test. Instead of fatal burns, the men died of asphyxiation. Courtesy of NASA.

USA." Their music swept over the crowd, many of whom were not American at all. Men in military uniforms from Israel occupied some of the seats in the audience, as did guests from the nation of India. It was an international audience to honor an international space shuttle crew.

John Clark was there as well. His job was to say the right thing. The *Columbia* families had asked him to speak for them. His wife, Laurel, was among those lost.

Clark stood at the podium with the Space Mirror behind him and asked if the people of America would be "space fearing or space faring." Other speakers talked about how the *Columbia* victims would be remembered in the same breath as the *Apollo 1* crew, lost in a launchpad fire in 1967, and the *Challenger* astronauts who perished in NASA's first shuttle disaster in 1986.

After the ceremony, Clark went from stalwart spokesman to grieving husband. It was his wife whose voice had beamed from orbit in January of 2003, talking about how the straps that held her equipment floated freely in the absence of gravity and how the metal buckles bumped into each other, jingling like a wind chime.[18]

John Clark was a flight surgeon at the Johnson Space Center in Houston at the time of his wife's mission, and he's seen NASA from the inside out. The day of the dedication at the astronaut Memorial, the agency's ears were still burning from the August release of the final report from the Columbia Accident Investigation Board.

Instead of giving unwavering support to the space agency, John Clark's attitude is one of skepticism. The agency may promise reforms, but Clark remembers the feeling of standing at the KSC landing site the morning *Columbia* was lost. He recalls what it was like to hear the radio messages of lost tire pressure blare over the loudspeakers, and then the loss of contact with the astronauts. At that moment, Clark thought the situation might mean the astronauts couldn't land and would have to bail out. An hour later, he learned the vehicle had broken apart and no rescue would have saved them. "I've learned hard lessons in aviation," he says. "That you have to stop and say, 'Wait a minute.' Things like Mission Completion Syndrome or get-there-itis can make you do things you wouldn't ordinarily have done before." Their son, Ian, still dreams of being an astronaut. He plays on spacecraft mock-ups in Houston and makes ships of his own in bed before going to sleep. Tougher questions may come later. Clark's concerns extend beyond his wife's fate and to the next crew to fly on the shuttle. "We want to make sure we do right by them," he says.[19]

When *Columbia* was lost and the Clark family was dealing with the painful days and weeks that followed, astronaut Jim Kelly was there for them. NASA assigned Kelly to act as the Clarks' casualty assistance officer. Now Kelly's safety would be on the line as a member of the first shuttle crew to fly following the accident.

Back on the Horse, Again

While NASA worked to implement the recommendations of the *Columbia* Accident Investigation Board, the International Space Station limped along in orbit. The space shuttle program's long-sought-after mission, to build and resupply the outpost, was on hold following the *Columbia* disaster, and its impact on the station was immediate. The standard crew complement of three people was trimmed to two in order to conserve supplies, and the steady stream of flights to deliver new modules and solar panels came to an abrupt halt.

Before the next launch, NASA would remove foam insulation from the external fuel tank at the spot where the nose of the shuttle was bolted on. Investigators were certain that this area, called the bi-pod ramp, was where foam tore off and damaged one of *Columbia*'s heat shields. Work to modify the external fuel tank was already under way when friends and family members of the lost shuttle crew gathered for the dedication ceremony at the Astronauts Memorial at the Kennedy Space Center in October.

One face missing from the crowd at the memorial ceremony that day was Margaret Conklin of Orlando.

She felt bad about that.

"Maybe I should have gone," says Conklin. "As a family member, I mean." The planned resumption of shuttle flights was still a year or so away, so the nervousness for her and of her family wouldn't start to escalate for months. Conklin's sister, astronaut Eileen Collins, would command the mission, and Margaret would be there, as she had for all of her blastoffs. Conklin knew from experience how bad a shuttle launch could be, since she was at the family viewing stand at Kennedy when Eileen launched aboard *Columbia* in 1999 and almost lost a main engine to a short circuit and a fuel leak. Margaret knew something was up. "My husband and a friend were standing next to me," says Conklin of the launch. "They said nothing was wrong, and I told them, yeah, there's something wrong. They're not supposed to be talking like that [over the radio]." There were many anxious days before Collins and her crew returned from that mission. Now this astronauts' extended family was making plans to gather in Florida to watch another launch, the first one to go since *Columbia*.

Back in 1988, two years after the *Challenger* disaster, Collins was visiting Margaret while on leave from the air force. It was during a morning chat in the kitchen of the Conklin home in suburban Orlando that Eileen announced she would apply to be an astronaut. That didn't go over well with Margaret, since she recalled standing outside her home in 1986 and watching *Challenger* explode moments after liftoff. "I remember seeing that," says Conklin. "I asked Eileen, how can you want to do that? She said, it's okay, it'll go again. There was this confidence about her. I mean, you have to be to get on that thing."[1]

Collins was a U.S. Air Force major at the time, and behind those all-American apple cheeks when she smiled there was a hard edge that hadn't gone unnoticed in Washington. Collins flew a C-141 aircraft on a rescue mission in Grenada during "Operation Urgent Fury" that earned her the Armed Forces Expeditionary Medal.[2] The shock of her announced ambition to become an astronaut was succeeded by greater shock when chief astronaut John Young telephoned. "She couldn't speak," Conklin says her sister later confided. "Then she asked, what would her job at NASA be, as a mission specialist or as a pilot?"

Pilot, Young responded.

"No woman had ever done that before," says Conklin.

After three successful missions on the shuttle, Collins's next flight was to have been to the International Space Station, following *Columbia*'s 2003 launch. When that ended in tragedy, Collins was instead charged with the return to flight that would define the final years of the shuttle program. NASA changed the flight's mission from delivering a new crew to the space station to conducting tests of safety equipment and procedures prompted by the shuttle accident.

A New Mission and a New shuttle

The new flight plan meant a crew shuffle. Astronaut Ed Lu and cosmonauts Yuri Malenchenko and Alexander Kaleri were supposed to fly with Collins and stay behind on the space station. Now they would have to wait. Lu and Malenchenko would board ISS, but they would get there aboard a Russian-built Soyuz spacecraft. Kaleri would be in a later crew. Original versions of the embroidered STS-114 crew patches featuring the names of the now discarded members became hot items among space collectors. The core shuttle crew remained the same. Collins was teamed on the flight deck with pilot Jim Kelly, and the two spacewalkers, Soichi Noguchi of Japan and Steve Robinson, retained their assignments. The extra loading and unloading of supplies meant more crew members were needed. Rookie Charles Camarda got the call along with two veteran astronauts with experience working aboard the older Russian space station Mir.

Crew member Wendy Lawrence has the distinction of having been turned down for a long-term flight aboard Mir because she wasn't tall enough. Mission planners in Moscow felt she couldn't perform a spacewalk in a Russian-built Orlan suit. During her next space mission, Lawrence commemorated that rejection by wearing a name tag on her blue flight suit that said "too short."[3] She did fly twice to Mir on crew exchange missions, one of which brought an astronaut who would also join the STS-114 crew to make the first post-*Columbia* flight. That was Australian Andy Thomas, the man who once raised the ire of Russian space officials when his Sokol spacesuit seemed too tight and needed adjusting after the space shuttle blasted off. Thomas's work on the shuttle mission would begin just minutes after launch. His first job would to be unstrap from his seat, float up to the cockpit, and start taking pictures of the external tank as it detached and floated away. The goal was to spot any divots in the tank's foam insulation that could indicate falling debris similar to what caused the *Columbia* disaster.

Thomas's next task would be the very next day, when he tried out a new device, the "boom." Shuttle crews couldn't see the belly of their spacecraft, even with cameras on the end of the fifty-foot-long Canadian-built robotic arm. The answer, built from scratch by the Canadian Space Agency, was a fifty-foot extension for the arm, called the boom. Thomas would grab it with the robotic arm to create a long spindly periscope, lean it over the side, and use the cameras and lasers on the far end to peek at the belly tiles. "Considering the complexity of what it does," said Thomas, "to interface with the shuttle, carry a suite of sensors, cameras, and lasers, and beam the signals back to the shuttle—if there was damage, we'd see it." The rookie who rounded out the STS-114 crew, Charlie Camarda, would assist with damage inspection and cargo transfer. He and spacewalkers Steve Robinson and Soichi Noguchi would also venture into new territory, testing ways to fix heat tile damage similar to that which caused the *Columbia* disaster.[4]

A big-picture question for this return-to-flight mission was which shuttle would be used. Collins's crew was supposed to fly on *Atlantis*; indeed, the shuttle was waiting to blast off when *Columbia* was lost. During the nearly two and a half years while the shuttle fleet was grounded, a long list of checks was ordered. The heat shields on the leading edges of all the orbiters' wings were removed, sent to the manufacturer, and scanned for defects. The spacecraft were inspected as well, and problems were found on the tail rudders. Each has a set of speed brakes designed to spread open like wings to slow the shuttle down. There was corrosion on both *Atlantis*'s brakes and *Discovery*'s. However, *Discovery* was farther along in a regularly scheduled maintenance period, and its speed brakes would be fixed faster. Therefore NASA pushed *Discovery* to the front of the line and told Collins's crew to use that vehicle.[5] It was a familiar craft for Collins, who had made her first flight aboard *Discovery*.

The shuttle would carry up two tons of cargo in an Italian-built warehouse module called Raffaello. The crew would grasp the compartment from its nesting spot in the cargo bay, lift it out with the shuttle's robotic arm, and snap it onto the station, where it would act as a temporary part of the outpost. The cargo in the compartment would come out and trash would go in. A growing problem on the International Space Station was that more stuff was going up on the Russian robot craft called Progress vehicles than was being brought back down. The unmanned Soyuz-like ships would dock on autopilot and drop off food, fuel, and equipment. But, unlike the shuttle, Progress craft can't land back on Earth.[6] They are simply allowed to burn up in the atmosphere. So while the shuttles were grounded, a lot of used gear was cluttering up ISS. "The logistics and the cargo transfer remained a big deal," said Collins. "More and more cargo was being taken up on Progresses, but not taken off. So the shuttle was needed to keep the station going." *Discovery*'s new crew members would help coordinate the flow. Wendy Lawrence was put in charge of offloading incoming cargo and the delicate process of loading trash back in. The Italian Space Agency built Raffaello and its two twin compartments, called Leonardo and Donatello, to act as portable warehouses nestled in the shuttle's cargo bay. Leonardo went first, during shuttle mission STS-100 in 2001, and mission managers discovered a critical problem early on. The cargo going into space was packed carefully so the weight was distributed evenly to keep the shuttle from flying off-balance. No one kept track of the weight of the objects coming back down before they were loaded. Those things were weightless in orbit, but they wouldn't be as the shuttle glided toward landing. Mission control had to call up the weight of each piece of gear or trash by radio.[7] The shuttle crew then used that information to strap the incoming cargo in just the right spot to distribute the weight.

Besides arranging the load in Raffaello on STS-114, it would be Wendy Lawrence's job to stock supplies for something called "safe haven." If the shuttle got damaged during liftoff, the astronauts would have the option of evacuating into the space station and staying there until a rescue shuttle could be dispatched. Lawrence spent time prior to liftoff figuring out what extra provisions the crew would need for an extended stay of a month or more. "The point was to minimize what we would need from the station to be self-sufficient," said the astronaut. "So we're taking a handful of items in Raffaello specifically for safe haven, like extra items for carbon dioxide removal [from the air], along with as much food as I'm allowed. Also, we'll take things for hygiene and clothing." Along with the extra supplies, the safe-haven scenario would mean the *Discovery* crew needed a way to get home. If their own shuttle was damaged and unable to make the dangerous reentry into the earth's atmosphere, they'd need to hitch a ride.

The Rescue Shuttle

On the day *Columbia* was lost, February 1, 2003, *Atlantis* was also ready to go. That shuttle was bolted to its external fuel tank and solid fuel boosters and was standing on top of a huge tractor called the crawler transporter. The crawler was a relic of Apollo days, when this same machine carried Saturn V rockets to the pad for voyages to the moon.

Atlantis's trip to the launchpad was cut short when NASA grounded the fleet after the *Columbia* disaster and sent engineers and astronauts alike scrambling to East Texas to look for wreckage. *Atlantis* was also the subject of speculation in the press. It was almost ready to go. If NASA had known the extent of *Columbia*'s damage, and feared that the shuttle couldn't survive the trip through the atmosphere, could *Atlantis* have made a rescue flight?

Columbia was gone, so the rescue idea was just the stuff of speculation—at least, early on. There would be no guesswork during *Discovery*'s return-to-flight mission. *Atlantis* stood poised for a rescue if it was needed.

Veteran astronaut Steve Lindsey would be in charge of the rescue shuttle in case Eileen Collins's crew ran into trouble and needed a ride back to Earth. This emergency scenario was made part of the preparation for Lindsey's post-*Columbia* flight, which would come right after Eileen Collins's mission. It was known as STS-121, but it had an alternate name, STS-300. *Atlantis* was ready when *Discovery* blasted off. If Collins's shuttle was damaged during takeoff and her crew was threatened, NASA would order the astronauts to evacuate into the International Space Station and wait for *Atlantis*. "They would use the combined resources of both the station and *Discovery* to keep themselves healthy," said Lindsey. "*Discovery* would stay docked, and once all the food and oxygen was used up, Mission Control would command the vehicle to undock and do a controlled burn [crash landing] into the Pacific." *Atlantis* would then dock.

Seven astronauts, including Lindsey, were assigned to their regular post-*Columbia* flight. The rescue mission would mean three of the crew would stay behind. Lindsey and pilot Mark Kelly (no relation to Jim Kelly) would fly the shuttle, and astronauts Piers Sellers and Mike Fossum would perform any spacewalks that were needed.

It would be a crowded flight back home, with the rescue crew and the once-stranded astronauts filling every inch of *Atlantis*'s crew cabin. The shuttle had never flown with more than eight people. A post-*Columbia* rescue landing would mean eleven people crowding the spacecraft. And seven of these people would be feeling the ill effects of a monthlong spaceflight, which would complicate the seating arrangements. "We're assuming that they would have been airborne for forty to sixty days," said Lindsey. "Like all station crew members, we'd have to fly them home recumbent or lying down, and we'd do the same thing for the 114 crew."

Spacewalk Plans Inspired by Home Depot

During Eileen Collins's shuttle mission to dock the vehicle to the International Space Station, astronauts Steve Robinson and Soichi Noguchi had originally planned to perform two routine spacewalks to install parts. That was before the *Columbia* disaster. Now their task was more challenging. Robinson and Noguchi would test ways to repair heat protection tiles that had been damaged like *Columbia*'s. Their work would be studied and copied by future shuttle crews. They would have to lead the way.

To repair damaged heat shields in situ might present difficulties even on Earth. Robinson and Noguchi needed to figure out what methods worked in the vacuum of space. Testing and brainstorming began long before they strapped themselves in for blastoff. The job entailed stocking the cargo bay with trays filled with broken test tiles. Once in orbit, Noguchi and Robinson would smear a repair compound onto the damaged tiles like peanut butter on bread. NASA's high-tech community gave the pinkish, caulklike substance a pair of low-tech nicknames. The astronauts called the stuff goop or goo. "The idea is to get the goo to adhere to the tile," says Robinson. "The catch is, when the tile is broken there's not much for the goo to stick to." Robinson and Noguchi quickly discovered just how difficult it was to work with goo/goop without gravity. The two boarded a NASA training jet that creates weightlessness by flying in a steep climb, then going downward in an arch so things and people float around the passenger cabin. The vehicle is called the "vomit comet" for the reaction rookie astronauts sometimes have if they eat prior to a training session. Floating inside the airplane's cabin during the brief periods of no gravity, Robinson and Noguchi tested squirting the repair compound, which looked like cake frosting. It was pretty messy, but better to encounter the problem here than in the vacuum of space, where goop on the fingers of a spacesuit glove could lead to a smeared helmet faceplate.[8]

Practicing with goop was only one piece of the puzzle. Finding the tools to help spread the material would take NASA to a Houston-area Home Depot do-it-yourself store. Agency engineers made notes and then built a series of palette knives and brushes so the spacewalkers could tamp down the caulking material while wearing a spacesuit with gloves as thick as a boxer's.[9]

Discovery's modified external fuel tank was delivered to the Kennedy Space Center in January of 2005. The plan was to bolt *Discovery* to the tank and roll it to the launchpad for a fueling test. The new tank had less foam in the place where NASA believed insulation tore from *Columbia*'s tank in 2003, causing the shuttle disaster. Launch managers wanted to pump a load of cryogenic fuel into *Discovery*'s new tank at the pad to see how it behaved. The test aroused concern that ice could form due to the cold fuel and perhaps break off in flight, damaging the shuttle. *Discovery* was rolled off the pad and given a new tank with an extra heater to melt any ice that appeared on launch day. NASA was taking no chances.

Launch Day

Blastoff was set for July 13, 2005, and the crush of reporters was the largest since Mercury pioneer John Glenn flew aboard *Discovery* in 1998. Close to three thousand journalists and technicians requested credentials to attend the blastoff. With security tight, most of the media were ordered to park in a baseball field off KSC property and take buses to the Press Site overlooking the launchpad. Network television crews set up tents for their anchors to use, while radio and print reporters occupied trailers or milled around the main press facility, nicknamed "the wedge" for its unusual shape.

Weather was the big worry, with thunderstorms forming close to the pad. If *Discovery* were to lose engine power minutes after liftoff, then an emergency landing back at Kennedy would be too dangerous because of poor visibility. The weather never had a chance to scrap the liftoff, though, because one of *Discovery*'s fuel gauges failed first.

The system uses eight sensors, called engine cutoff or ECO sensors, each about half the size of a pack of cigarettes. They're buried inside the butterscotch-colored external fuel tank in an environment where no person could live, surrounded by hydrogen and oxygen chilled as low as 400 degrees below zero. The sensors' job is to tell the shuttle's main engines when gas is running low. The engine's turbo pumps generate 70,000 horsepower during the eight-and-a-half minute trip to orbit. NASA's nightmare scenario is having the tank run dry with the sensors not working. If the pumps were running without fuel, the engines might rev out of control and explode. Not all eight sensors were working on *Discovery*'s launch day, so the countdown was stopped. "All we can say is, shucks," said NASA shuttle program manager Wayne Hale. The reaction of the press was less understanding.[10]

Fewer reporters came back for the rescheduled liftoff. MSNBC's set was left abandoned, offering photographers a better perch for the next attempt. The agency's new administrator had a different constituency to deal with following *Discovery*'s fuel sensor failure. Michael Griffin had been entertaining a roomful of power brokers from Congress. The postponement meant this audience, including U.S. senators Kay Bailey Hutchinson and Bill Nelson, former presidential candidate John Kerry, and House majority leader Tom DeLay, went home disappointed. Senator Nelson flew on the shuttle in 1986 just weeks before the *Challenger* disaster, and his tone was conciliatory. "When my flight went, we held the record of scrubs with four," he said. "We finally went on the fifth try almost a month later." That offered little solace with NASA's future as uncertain as ever.

Liftoff finally occurred on July 27 on a blisteringly hot day at the Kennedy Space Center. Engineers never found the precise cause of the sensor failure. Launch controllers pumped propellant into the 154-foot-tall external fuel tank and waited to see if the problem would occur again. It didn't. The main engines ignited, followed six seconds later by the roar of the solid fuel boosters, and NASA's return-to-flight mission was under way.

Hundreds of cameras were trained on the spacecraft as it soared to orbit. Ground-based cameras followed *Discovery*'s progress, as did NASA jets equipped with more photographic equipment. Perhaps most dramatic were the cameras inside the external tank itself, as well as video equipment inside the booster rockets, and the astronauts' handheld cameras.

Figure 47. The protuberance air load ramp, or PAL ramp, on space shuttle *Discovery*'s external fuel tank, where a two-foot section of foam insulation tore loose during liftoff. The foam problem forced NASA to postpone all shuttle launches during *Discovery*'s return-to-flight mission. Courtesy of NASA.

That's when evidence of more falling foam insulation was discovered.

NASA shuttle managers Bill Parsons and Wayne Hale appeared sheepish as they faced the press with news that the shuttle fleet had been grounded again by the now open safety issue. "Until this is closed," said Parsons, "we will not fly again, so we may as well let that out. Closing an open issue like this takes a lot of work."

The foam on the midsection of the external tank was an area NASA had thoroughly examined and cleared for flight on *Discovery*. "We were wrong," declared Parsons. In one spot the foam forms a shield called the protuberance air load ramp, or PAL ramp for short. This protects wiring and propellant lines running along the outside of the tank from air rushing over its exterior during the trip to space. Fixing the problem could be time-consuming, and shuttle *Atlantis* was waiting in the wings. Just stripping the suspect foam off might not be enough. And the only alternative to leaving portions of the tank unprotected would be a lengthy redesign effort.[11]

The PAL ramp wasn't the end of *Discovery*'s problems. Early photos by the robotic arm's boom extension cameras, later confirmed by snapshots taken by the space station crew out the windows of the orbiting outpost, showed something near the doors of the shuttle's nose landing gear. Heat-resistant cloth stuffed between tiles had popped loose in two different spots. "We just don't know enough about this to feel good about it," said NASA's Wayne Hale. And if NASA didn't feel good, something would have to be done. Hale called it the "new NASA."

Before the 2005 return-to-flight mission, no astronaut had ever ventured near the shuttle's belly while the spacecraft was in orbit. It was too dangerous. The black tiles were all that stood between the crew and incineration during the fiery trip back through the atmosphere at mission's end. *Columbia* demonstrated how bad a reentry could be with a damaged heat shield. The *Discovery* astronauts kept a photo of the lost shuttle crew taped to the wall of the flight deck. With the faces of those former colleagues staring out, *Discovery* spacewalker Steve Robinson heard Houston tell him to suit up, perch on the end of the space station's robotic arm, and

Figure 48. Space shuttle *Discovery* photographed from inside the International Space Station. Closer inspection showed gap filler sticking out between tiles around the nose landing gear. One piece is visible below and to the right of the rectangular doors. Courtesy of NASA.

repair the heat shields. He would carry a pair of forceps and a modified hacksaw to pluck or cut the material, called gap filler, from between the tiles.

It was a spacewalk like no other. Usually crewmates travel in pairs outside the shuttle, and often around the comfort of the payload bay. If a problem occurred like a broken safety line or a suit puncture, the second spacewalker would be there. Robinson would face this job alone. Japanese astronaut Soichi Noguchi would venture out of the airlock with his crewmate, but Robinson would go by himself to the underside of *Discovery*'s nose. "You don't want two crewmen on the [space station] arm," says NASA Spacewalker coordinator Cindy Begley. "It causes too much bouncing, which could damage a tile." Once he was moved to the work site, the spacewalker faced a view not unlike what navy divers see while they work outside a submarine. It looked like acres of blackness, punctuated only by the white serial numbers stenciled on the tiles to tell one from the other. A lot of people in Mission Control didn't like the repair idea, and the astronauts were worried as well. "When we heard the plan from Houston, some of us had misgivings," said *Discovery* crew member Andy Thomas. "We knew the implications."

The notion of putting a spacewalker close to the belly tiles was something Steve Robinson and Soichi Noguchi had considered, even though their mission didn't call for it. Spacewalkers had already amassed a lot of experience working outside the shuttle. Astronauts had floated off away from the payload bay to work on remote locations on the International Space Station, or used jet packs to nab wayward satellites. But no one had ever tried to go over the side of the shuttle's payload bay and venture to the delicate black belly tiles that would take the brunt of the heat of reentry. A damaged tile in that location could spell disaster.

Before *Discovery*'s launch, mission planners thought about a new way to get people there, and some of the ideas were wild. One was a tripod framework the spacewalkers would use as a brace. "At first, we thought it would be a good idea to have something to stabilize our bodies like a lunar lander," said Noguchi. "We would maneuver ourselves with a jet pack and come up against the belly. It sounds nice, but after a couple of engineering evaluations, we felt the tripod lander wasn't a good technique." The two spacewalkers began the return-to-flight mission with no good way to visit the belly. They'd help invent one along the way.

Robinson would perch on the space station's robot arm and inch toward the underside of the nose. Before he ventured outside, spacewalk organizers on the ground took heat tile samples and the tools that Robinson would have with him to see what might go wrong. Tests showed he was unlikely to scratch or chip a tile if he sliced the gap filler off with a custombuilt hacksaw. The next nightmare scenario was Robinson accidentally prying a tile off. But technicians felt that the glue used to paste the tiles on the spacecraft was too strong for that.

While scratching or loosening a tile might be avoidable, head-butting one was still a concern. "The spacesuit is huge," said Robinson. "You have to be very aware of your body mass. I'm going to be very concerned about my helmet as I lean forward toward the shuttle." The day of the spacewalk, Robinson and Noguchi donned their heavy suits while pilot Jim Kelly and mission specialist Wendy Lawrence floated into the station's Destiny lab. It was their job to drive the fifty-eight-foot-long robotic arm with Robinson perched on the end. A small platform called a portable foot restraint would attach to the arm, and Robinson's boots would clamp into it. Radio contact would be limited, since the shuttle might block the radio signal from Robinson's suit. Mission Control's remedy was for the spacewalker to keep talking. If the arm operators didn't hear Robinson's continuous commentary, they were supposed to stop the arm's movements immediately.

The hours of testing on the ground and the care that went into customizing his tools inside the shuttle would lead to an operation that took only seconds. Robinson arrived at the work site, reached out with his gloved hands, and plucked out each of the fillers with little trouble. The hacksaw and forceps stayed neatly in his tool bag.[12]

Discovery landed safely, but members of the mission management team were left somewhat haggard by the string of problems on the return-to-flight mission. There had been the foam loss from the external tank, the damage to the gap fillers requiring repair, and finally a torn heat protection blanket just below the black-rimmed cockpit windshields where Eileen Collins looked out during the glide to the ground. The blanket almost prompted another repair spacewalk, but NASA opted against it. The decision was made despite a worst-case scenario that had the twenty-inch sheet of fiberglass fabric tearing off during the supersonic reentry into the atmosphere and punching a hole in the braking flaps on the shuttle's tail rudder.

During their first comments to the press, the *Discovery* crew admitted that the fate of *Columbia* was on their minds during reentry. "There was some trepidation as Eileen pressed the button to begin the deorbit burn," said pilot Jim Kelly. Collins's feeling was "We're going to get through this." Bad weather kept the crew from landing in Florida. *Discovery* headed to California for an early-morning touchdown. The path from orbit would have taken the shuttle over downtown Los Angeles, so NASA tweaked that course to avoid the city. The image of wreckage raining down on East Texas in 2003 was something the agency didn't want to see repeated in the City of Angels if *Discovery* were to break up on final approach.[13]

When *Discovery* had made it back safely, NASA opted not to move *Atlantis* up for the next flight, but to send *Discovery* again for the second post-*Columbia* test flight. *Atlantis* would be held in reserve for the launch after that, to carry up new solar panels and a section of the International Space Station's spinelike truss.

After waiting in the wings, Steve Lindsey and his crew would be tasked with a flight as complex as the first post-*Columbia* flight, if not more so. They would pick up where Collins and her crew left off, but where Steve Robinson plucked gap fillers from the underside of *Discovery*'s nose with his feet clamped to the space station's robotic arm, Lindsey's spacewalkers would do something more acrobatic. They would perch on the end of the fifty-foot inspection boom, which would be in the grasp of the shuttle's robot arm. "We wanted to take the boom and put a man on the far end," says Lindsey. "We wanted to try out its ability to use it as a repair platform." Both Collins and Lindsey were under pressure to maximize the inspection and repair practice. NASA was clearly interested in resuming the construction of the International Space Station as soon as possible. However, the aging shuttle fleet wouldn't make that task easy.

Being the veteran of three successful shuttle flights didn't make the family meeting at Lindsey's house after the *Columbia* accident any easier. "I waited until way after the accident to talk about it," he says. "No tough questions. They just wanted to know what happened." The four-time space veteran had had a heart-to-heart with his wife and children about the hazards of space travel long before *Columbia*. They watched from the Kennedy Space Center as he blasted off on his first flight. Next there was the mission featuring John Glenn. Lindsey was then assigned his first shuttle mission as commander, on a flight to the International Space Station to deliver the U.S.-built airlock nicknamed Quest. Now he was in line to make the second mission following *Columbia*. "I've assured them if a time came that I wasn't confident that we were doing everything we could to make it as safe as practical, I wouldn't want to be flying it."[14]

There would be no need for Lindsey to make the rescue flight to help *Discovery's* crew.

As NASA resumed shuttle flights and tried to restart its long-sought-after mission to build the space station, mission management team leader Wayne Hale was philosophical. The first post-*Columbia* flight, led by Eileen Collins, marked the beginning of the end of NASA's space shuttle program. During that critical mission, Hale had his eye partly on the safe return of the astronauts, but also on the next generation of spacecraft designers who would take over in the future. "Learn from what we did and what the designers of the space shuttle got right, and from what they did that was not so right," he said. "And when you build the next generation of spacecraft, take what is good and build on that, and avoid what didn't turn out so well."[15] It's not just shuttle engineers who may be thinking that they're in the twilight of their careers. The next generation of managers, engineers, and astronauts will likely face some familiar problems as NASA pursues its new mission to go back to the moon and on to Mars. There will also be a whole new collection of problems to solve, both technical and human.

The Moon, Maybe, but Mars?

During the lifetime of the space shuttle, visitors to Central Florida's Atlantic coast would park their cars and motor homes at choice spots along U.S. 1 to watch *Columbia, Challenger, Atlantis, Discovery,* and *Endeavour* leap from their launchpads and fly to space on a column of fire. Now the end of the shuttle program means that the best these space enthusiasts will be able to do in the future is visit the orbiters in museums. When America abandoned its lunar program in 1972, leftover Saturn rockets and test vehicles from the Apollo moon flights were doled out in this fashion. Some got places of honor, while others languished in obscurity.

After its failed flight to the moon, the *Apollo 13* command module called Odyssey spent years in the lobby of an airport in Paris, while the more desirable Columbia capsule from the *Apollo 11* mission sat on display at the Smithsonian Institution's National Air and Space Museum in Washington, D.C.[1] *Atlantis* might go there as the first space shuttle to dock with the Russian space station Mir. *Endeavour* might be chosen, since it carried the three spacewalkers who rescued the Intelsat 6 satellite by grabbing it with their gloved hands. The choice might go to *Discovery*, which was launched as the return-to-flight vehicle after both the *Challenger* accident in 1986 and the loss of *Columbia* in 2003.

As visitors to the Smithsonian relive the moments of the shuttle program and Apollo, NASA will be looking to the Apollo days as the starting point for its own move back to the moon as it sets its sights on Mars. The retirement of the shuttle fleet and the beginning of the focused mission of Project Constellation won't be as easy as simply shifting gears, and there's a lot at stake.

Once NASA grounds its winged orbiters permanently, there's no going back, and the agency will be expected to make the moon-and-Mars program work. Scientists and researchers who are studying the challenges of returning people to the moon say there will be new demands on the next generation of spacecraft and astronauts, and on the agency itself. If NASA fulfills its job of actually sending crews to the planet Mars, the agency will be dealing with an even more alien and hostile environment than it encountered during the early moon flights of Apollo.

Apollo Veterans Say the Moon Isn't as Scary Now

According to the men and women who made Project Apollo a success, America won't have to deal with the mysteries and uncertainties that plagued the first moon landing. Project Apollo was a huge undertaking for the United States in the 1960s, and its supporters proudly point out that it was achieved without laptops or personal computers, but rather by men and women armed with slide rules. The program's goals were deceptively simple: get to the moon safely, grab some moon rocks, set up some experiments, drive around in a lunar buggy, and return safely. That was easy to say after *Apollo 17* and the last lunar landing. For the first touchdown, by *Apollo 11* three years earlier, there were a lot of unanswered questions. Images of Neil Armstrong and Buzz Aldrin bounding around the Sea of Tranquility seem quaint and even mundane now. But in the minutes prior to landing, NASA was extremely nervous about what might happen during touchdown on the lunar surface.[2]

The astronauts spent eight hours walking around the moon's surface and returned to a hero's welcome—along with a month in quarantine in case any unknown moon microbes hitched a ride back to Earth.

Figure 49. Astronaut Neil Armstrong peering back at the cameraman during this training session in a lunar module simulator at the Kennedy Space Center in Florida prior to the launch of *Apollo 11* on the world's first manned moon landing. Courtesy of NASA.

Three other Apollo crews landed successfully before John Young and rookie Charlie Duke touched down in the lunar module Orion on the Descartes Plains of the moon. By then, flight crews say, they just hit the ground running on the gray, cratered surface, with nothing being particularly scary or alien. But for future long-duration moon missions, there are disquieting lessons from Apollo.

NASA Plans to Go from Short Stays to Long-Term Moon Missions

The most obvious concern is the spacesuits. The next Apollo crew to fly after Young's would be the last. *Apollo 17*'s moonwalkers Gene Cernan and Harrison Schmitt spent more time on the surface than any previous crew, and lunar dust got caked in the joints of their spacesuits. Astronaut Carl Walz believes that, after three moonwalks, contamination in the places where the gloves and helmets snapped on made a fourth venture outside impossible. "We need to develop suits that mitigate that dust problem," Walz says. "In fact, the suit design itself was a problem. It made you walk in an unnatural fashion, and that kicked up more dust. So one of the things we need to do is develop a suit that's more flexible and reduces the mass."

Mission designers for the Crew Exploration Vehicle recognize that people on a lunar base will have to survive for longer periods without regular visits from Earth. NASA often quantifies the cost of space travel by looking at it "per pound." In other words, what's the cost in fuel and ground support to put a pound of anything into orbit? That price tag for shuttle payloads is around three thousand dollars a pound. Mission planners looking at the bottom line are drawing inspiration from expeditions on Earth like those to the Arctic and studying how those trips make the most of limited supplies.

One of the papers delivered at the 1st Space Exploration Conference in Orlando was from Patricia Downey of the Bechtel Corporation. It dealt with the basics of building a moon base. Astronauts on long moon assignments will have to be resourceful, and their equipment must be versatile and built to last. "There's no Home Depot to get spare parts, right? I've worked in really remote areas, and you want something right there that's recyclable," Downey says. "Something that's modular that you can reconfigure into other things to do other jobs. Otherwise, you might have to wait weeks or even months for a shipment to arrive." Downey is painting a picture of astronauts using simple tools on the moon that can be broken down into basic pieces and rebuilt into other things. This experience is meant to be a test bed for making trips farther away.[3]

All of this debate may be leaving NASA somewhat bemused, since the agency is still settling on how to get back to the moon in the first place, to say nothing of the intimidating trip to Mars.

Building the CEV and selecting crews are just two issues that mission planners are wrestling with. What to do when the astronauts aboard the next-generation spacecraft arrive on Mars is another.[4] They know that building a moon base will seem like a summer stroll compared to the challenge of putting people on the Red Planet.

Dealing with the Dust and the Distance of a Trip to the Surface of Mars

No human has ever set foot on Mars, but one group of scientists has come close, and that experience taught them firsthand how difficult it is to get around up there. They spent more than a year exploring the planet vicariously through a pair of $400 million remote-controlled robots called Mars Exploration Rovers. The golf-cart-sized machines were each equipped with twin cameras perched on a short mast, as well as a small drill to poke and prod interesting rocks. The rovers also used a spectrometer to determine what Martian rocks are made of. Shortly after liftoff, the two robots were given their more familiar nicknames, Spirit and Opportunity. Mission managers learned practical things along the way, like how to avoid toppling their rovers into craters, getting stuck in sand dunes, or being fried by solar storms.[5] They believe these lessons are applicable to future missions featuring astronauts.

"Dust is going to be everywhere," says Mars Rover principal investigator Dr. Steven Squyres. "There will be incredibly fine grains of sand clinging everywhere, into your spacesuit, in your food, just everywhere. People there are really going to get tired of the dust."

More than a nuisance, the Mars dust threatened the ability of the rovers to do their jobs. Each robot has wing-shaped solar energy panels to generate electricity, giving the Spirit and Opportunity rovers the appearance of wheeled beetles. During dust storms, the fine grains would settle on the solar panels, screening the amount of sunlight that could fall on the panels to create power. This same dust problem could degrade the joints, seals, and lubricants that human beings will depend on during long stays on Mars. There's also a possible biological hazard from the reddish dust that NASA is still studying. "You better have good weather reports," Squyres adds. "Astronauts could run into trouble in a hurry if a dust storm hits or if there's a solar storm with lots of radiation hitting the planet. There's no protective Van Allen belt around Mars like around Earth. You better have notice to seek shelter."

Steve Squyres's quick wit has made him a favorite with the press at question-and-answer sessions. While some mission scientists describe their work dryly and accurately, Squyres will talk about how the Spirit rover rolled up to a rock. "We threw everything in the tool bag at it," he'll say. Squyres is always good for a quote. His office in the Department of Astronomy at Cornell University in Ithaca, New York, has no pictures of Mars on the walls. New ones came in from the rovers all the time during their mission, so Squyres just downloaded the new ones to his laptop. His wall space was reserved for pictures by his kids.

In loading astronauts onto the Crew Exploration Vehicle and sending them to Mars, perhaps the biggest hazard is the distance. A one-way trip from Earth to the Red Planet could take up to a year. "If you get into trouble on the moon, you just hop in your rocket and come home," says Squyres. "With Mars, it's harder because of the length of the supply chain. The isolation is going to be a big factor."

Along with practical advice, the Mars Rover team also have a good idea of what people might see if they were standing on the surface of the Red Planet next to one of the rovers. Spirit landed inside Gusev Crater, which is thought to be an ancient lake bed. Opportunity touched down on the Meridiani Planum, where scientists believed a shallow salty ocean once existed. "On Meridiani, you'd find sand dunes and knee-high rifts in some places," says Squyres. "There would be some sinkage into the sand, and when you get to a crater, there would be exposed rock. Like Endurance Crater—very steep vertical or overhanging ledges." The point is that Mars isn't the same all over, and the conditions and dangers would be very different depending on where the astronauts were. At Spirit's landing site at Gusev Crater, the rover rolled up onto a series of mountains that mission scientists dubbed the Columbia Hills, after the lost space shuttle astronauts. Photos from the rover show what future crews might see, including the view down onto the rusty Martian plain with a butterscotch sky in the background. "Walking around there would be like strolling the dry valleys in Antarctica," says Squyres. "It's a mixture of rocks and sand, a very fine-grained sand. There wouldn't be a lot of slipping and sliding around." Looking down onto Gusev Crater, these astronauts might also see the many dust devils that twirl around the surface, reaching heights as great as one hundred meters. These whirlwinds may look dangerous, but data from the rovers indicate they wouldn't be like tornadoes on Earth. "Despite the fact that the winds get quite high," says Squyres, "the atmospheric pressure is so low that it could only lift Martian dust grains. It's not strong enough to pick up a person. You'd see something like cigarette smoke with very fine-grained sand."

The rover experience also taught mission scientists eye-opening lessons on how tough a place Mars really is. Both of the landing sites were believed to include sources of water long ago, which led to press accounts of how Mars might once have been hospitable to life. Mission scientists think otherwise now.

The water that might have existed at Meridiani long ago was ankle deep in the shallows and knee deep at the most. That Martian ocean is believed to have been very salty and probably acidic. The notion that life might have existed on Mars at one time may not work at all. Proponents of past life as we know it on Mars point to life on Earth that can exist in conditions ranging from solid ice in Antarctica to volcanic vents under the sea to the acid of car batteries. Other scientists believe that life has to form in favorable conditions before it can move to more difficult spots and thrive. Mars may have been so rough a place that no nice places existed to give life a chance.[6]

The Psychological Impact of Living and Working on Mars

That leaves questions about the barren and lifeless conditions that future astronauts will face on Mars, where the scenery will never change from day to day. "What will living on Mars be like at the gut level, the human level?" asks Squyres. "How desolate it is. How would you feel as a human to immerse yourself in that?" The Mars Rover scientist isn't alone in his concern over the psychological impact of a long stay on Mars, not to mention the arduous trip needed to get there.

NASA's first trip to Mars will be the first time human beings have existed in a place where they can't see the earth easily. During the Apollo moon landings, the Blue Planet was dramatically visible, and that meant home was comfortingly close. On Mars, the earth will appear as a tiny dot in the sky, and NASA researchers are concerned about that. "If you're on Mars and the earth isn't there, with all your family, friends, and culture, we don't know what the psychological impact of that will be," says Dr. Nick Kanas, principal investigator in a series of NASA studies on the effects of long-term spaceflight. The retirement of the space shuttle and the planned introduction of the Crew Exploration Vehicle are prompting the space agency to study who is best suited to make a moon or Mars flight.

Common sense might dictate that NASA's newest and least experienced astronaut candidates would be the last ones chosen. Nick Kanas disagrees, and he points to the changing psychological demands of long space trips. "I think that's because the missions have changed from the sitting-in-the-can flights in capsules to doing the long-haul missions. People are coming into the (astronaut) corps with longer missions in mind." NASA observers say the kinds of people the agency will be looking for to make moon or Mars trips probably won't change. Applicants are currently required to have college degrees in math or science and to pass rigorous physical and psychological exams. Pilots who fly the spacecraft typically have significant flight experience in high-performance aircraft, mostly in the military.

Despite the long lines of American and international candidates waiting to go on the space shuttle, NASA may face problems in finding people willing to go to Mars. A round trip to Mars could take two and a half years, and many potential crew members will think long and hard about leaving their families for that long. It could be even tougher to get people to volunteer for the very first Mars mission, which will probably loop around the Red Planet but not land on its surface. Astronauts may be willing to give up two years of their lives to walk on Mars, but not to fly close and then turn back.

Given the length of a Mars mission, researchers like Nick Kanas are carefully studying how people react to the isolation of the International

Space Station and the Russian space station Mir. The results of these studies point to potential problems on Mars and who might cope best. "The crew will have to be more independent," says Dr. Kanas. "It will take twenty minutes for questions sent to Mission Control on Earth to get an answer, so they'll have to take care of problems on their own. If there's a medical emergency or someone gets out of control and there's a suicide attempt, they'll have to deal with it." That sense of independence may include knowing when an astronaut crew should rely on technology and when they shouldn't. Computer breakdowns on the International Space Station caused major problems for the first crew members. In an emergency, though, they could abandon the outpost and be back on Earth the same day. Mars crews won't have that option.

Independent thinking among astronauts could also mean a jarring change for NASA and Mission Control, which is used to having a tight leash on activities in space. In addition, there will be less contact between the astronauts and their friends and families. The occasional gift packages that space station crew members received from home during visits by the space shuttle won't be arriving when people are on Mars. It's just too far away.

The length of Mars flights may also influence NASA's selection of crews. Each flight will likely be led by a commander and a pilot, just as on the space shuttle. The remainder of the crew complement would include experts in engineering, medicine, geology, and psychology. But the mix among the crew members is something NASA is studying closely. "Seeing the same six or seven people over two years may be tough," say Dr. Kanas. "The same stories coming from one crewman or another may be fun and interesting for now, but it may get boring after a while."

Psychologists say the right combination of people is the key to a successful long-term Mars mission. The first rule of thumb may be, mix it up. Their message is, don't send only one woman or one non-American crew member, to avoid isolating anyone within the group. A notable exception was the second long-duration mission featuring Americans and Russians aboard the Mir space station. Shannon Lucid teamed up with two male cosmonauts, Yuri Onufriyenko and Yuri Usachyev. Researchers say they got along fine. Despite the favorable result of that Mir mission, NASA is looking for ways to avoid friction. One idea is putting future Mars crews together on a space station or an outpost in an isolated spot, like Antarctica, to see how they get along for long periods of time.[7]

When proponents of NASA's new quest to send people to Mars speak on the subject, they often talk about how the flights are years away, and schoolkids currently studying the subject will grow up to be the crew members that will go. The point was recently brought home to NASA's most experienced astronaut. Space veteran John Young, who walked on the moon during *Apollo 16* and commanded the first launch of the space shuttle, had attended Princeton Elementary during his formative years in Orlando, and years later he went back for what he thought would be a routine school visit. Kids often ask astronauts things like how they go to the bathroom in space. That didn't happen during John Young's visit. The fifth graders wanted to talk about going to Mars, and they wanted specific details. "These kids were smart!" says Young. "They didn't want to talk about being an astronaut, they wanted an engineering report on going to Mars." When the Apollo veteran was that age, he reflected, he occupied his time with model airplanes.

If that brand of youthful speculation gives way to actual progress, and the pieces come together for trips to the moon and perhaps Mars, NASA would do well to remember what the lack of a mission did to hurt the space shuttle program and how the wrong perception of the mission of the Apollo moon program did damage as well.

Epilogue

Not everyone gets a second chance in life, but NASA may be able to resolve its past mistakes and move away from the troubled space shuttle program. A new chapter is in the making with a true mission to fulfill, assuming there's the national resolve to accomplish it. Still, there's a lot that can happen before NASA returns to the moon and heads to Mars. These what-if scenarios could put the agency in the same compromised position it occupied during the days of the space shuttle. One possible problem is another shuttle disaster before the program is retired.

NASA has only three of the original five spacecraft remaining in its fleet, *Discovery*, *Endeavour*, and *Atlantis*. A launchpad explosion at the start of a mission, or a malfunctioning landing gear during the perilous landing, could mean another spectacular accident. *Challenger's* disintegration in a fireball shortly after liftoff demonstrated that even successfully leaving the launchpad is no guarantee of safety. And the threat of being hit by space debris, wreckage of all sizes left circling the earth by previous launches, makes even the benign week or so the astronauts spend in orbit around the earth potentially hazardous.

If another shuttle accident occurred, and the names of more lost crew members were etched into the Astronauts Memorial at the Kennedy Space Center, it could shake the determination of Congress and the nation to keep launching shuttles to complete the International Space Station. A premature end to the shuttle program would complicate NASA's effort to keep its current workforce of engineers on the payroll so that at least some of these specialists can be utilized to build and launch the new Crew Exploration Vehicle.

Irreparable damage to the space station itself could be another complication. The fire aboard the Russian space station Mir in 1997 demonstrated how dangerous life can be aboard a space outpost. The message was driven home later that same year by a runaway Russian cargo craft that struck one of Mir's compartments, piercing the hull and allowing the air to rush out. The crew aboard Mir survived both incidents, but the mishaps point out that no spacecraft is immune to lethal damage, even the International Space Station. The premature end of the station would give NASA no reason to continue flying the shuttle, again hastening the retirement of the fleet. If station operation were suddenly halted or cut back, perhaps the most immediate impact would be on NASA's credibility to pursue international projects in the future.

Russia has participated extensively in the space station by contributing the boxcar-sized Zarya and Zvezda crew compartments and the Pirs docking module, along with visits by numerous manned Soyuz spaceships and unmanned Progress cargo vehicles. Canada has also been represented on the outpost by its Space Station Remote Manipulator System, the robotic arm the crew members use to grasp objects outside the station.

Europe and Japan might be the big losers if the space station isn't completed, or is closed down early by the sudden cancellation of the shuttle program. The European Space Agency has invested heavily in developing a laboratory compartment for the station called Columbus, and the Japanese Space Agency, JAXA, has done likewise with its lab module called Kibo. Those compartments spent years on the ground waiting to be launched before the *Columbia* disaster and after the shuttle broke apart during reentry in 2003.

If NASA successfully completes the station and retires its space shuttles, it will be free to pursue its moon flights and learn from the successes and mistakes of the original Apollo voyages in the 1960s.

The Pitfalls of the Apollo Moon Program

Project Apollo, which made names like Armstrong and Aldrin part of the national vocabulary, had a specific mission that the shuttle lacked. The downside was that there wasn't a long-term strategy to keep the moon program going after the initial success of the *Apollo 11* lunar landing. Losing the total potential of Apollo always bothered astronaut Pete Conrad, who commanded NASA's second moon landing. "After *Apollo 11*, we sort of fell off a cliff," he reflected. "Everyone put their hands up and said, 'What now?'" Even NASA had played into the goal, set by President John Kennedy, of putting a man on the lunar surface by 1970 and before the Soviet Union. That short-term goal turned against the space program, even as the ticker tape fluttered down on moonwalkers Armstrong and Aldrin.

Conrad reminisced about his Apollo days with the same gap-toothed grin and corrosive wit that was apparent in his first long-duration Gemini mission with astronaut Gordon Cooper in 1965 and his arrival at the Ocean of Storms on the moon during the *Apollo 12* landing in 1969. When he and lunar module pilot Alan Bean touched down following the *Apollo 11* mission featuring Armstrong, Aldrin, and Collins, it was clear to Conrad that the Apollo dream was crumbling. The big casualty was something called Apollo Applications. "We were talking about a space station by 1970, and a moon base after that," said Conrad. "After that, you're talking Mars. By 1983 we would have been there." None of that happened, and Conrad points to the man for whom the Johnson Space Center in Houston is named. "What you and the public didn't hear was that President Johnson had Vietnam and the Great Society to pay for," said Conrad. "He brought us in [to the White House] back in 1967 and told us, 'You stay on schedule for the moon, but you knock off this way-out stuff,'" meaning Apollo Applications.[1] Conrad wouldn't live to hear President George W. Bush's plan for a possible Mars flight through his new initiative called Project Constellation. A motorcycle accident would claim the astronaut's life in 1999.

Figure 50. *Apollo 11* astronauts Mike Collins, Neil Armstrong, and Buzz Aldrin outside a lunar module simulator at the Kennedy Space Center. Courtesy of NASA.

The challenge for NASA, as it moves from the era of the space shuttle to a planned return to the moon and maybe a trip to Mars, is to learn from both the good and the bad results of Apollo. It was competition with the Russians that helped fuel the American lunar landings. A new space race could provide an additional complication for the U.S. space program as it contemplates a return to the moon. That competitor might be the People's Republic of China.

"One Small Step" to Save the Apollo Landing Sites

The Chinese space program has been referred to as the last secret space effort on earth. The communist nation successfully launched one of its Soyuz-derived Shenzou space capsules in October of 2004, putting crewman Yang Liwei in the same league as astronaut Alan Shepard and cosmonaut Yuri Gagarin. All three men were the first of their nation to go into space. But don't refer to Yang as an astronaut. He and his space-faring colleagues in China go by the name *taikonaut*, derived from the Chinese phrase *tai kong* or "outer space." The communist nation is hinting at building its own space station, and possibly mounting a manned mission to the moon around the time NASA intends to go back.[2]

That possibility could make the small faction of lunar preservationists concerned. These people don't work for NASA, and they're not involved in the effort to send people back to the moon. They're laboring to safeguard the places where the original Apollo crews explored the lunar surface. The thought of people, Chinese or otherwise, tromping around *Apollo 11*'s landing spot at Tranquility Base someday has this small but vocal group lobbying NASA and Washington to take action. The lunar surface that astronaut Buzz Aldrin once described as "magnificent desolation" is seen by some archaeologists as a historic treasure trove, and they cringe at the notion of Apollo artifacts being scooped up by profit hunters. "It's like the Liberty Bell," says Dr. Beth O'Leary, a professor of anthropology at New Mexico State University. "You can see a picture of the Liberty Bell, but to actually see it in place is integral to its meaning." She counts herself among the handful of people who are working to get the Apollo landing sites protected under federal or international law.

When *Apollo 11* astronaut Neil Armstrong set foot on the moon, one of the first things he did was take a small bag from a pocket of his spacesuit and snatch a quick bagful of lunar soil. This "contingency sample" was gathered in case he and crewmate Buzz Aldrin had to make a quick departure from the moon. NASA wanted something to show for all the time and expense of the lunar landing. The crew later took precautions to make sure their departure went smoothly. They tossed overboard anything they didn't need, to lighten the load prior to liftoff from the moon. "There's a scatter zone around the *Apollo 11* landing site," says O'Leary. "We have the space capsules in the Smithsonian, and we don't let people walk up and carve their initials into them, so why not protect the things that are up there [on the moon]?" According to an inventory she assembled from NASA documents, Armstrong and Aldrin left more than a hundred items littering Tranquility Base. The list includes food items still wrapped in clear plastic bags, two complete spacesuits, a Hasselblad camera, and scoops that were used to gather lunar samples. There were also items the crew was supposed to leave, like the U.S. flag they planted in the soil, a silicon disk with greetings from four living U.S. presidents and seventy-three other heads of state, and a cloth patch and medals to remember the *Apollo 1* crew who died in a launchpad fire and two dead Russian cosmonauts. "People could buy them or sell them and they'd disappear," says O'Leary. "And the only people who'd see them are people who can afford it."[3]

Lunar preservationists may have reason to be concerned. Space collectibles are big business, and entrepreneurs inside and outside the space program are working to cash in on the public's interest in artifacts certified as having flown in space. Right now, most buyers and sellers are limited to acquiring flown flags and other small items. Cosmonauts helped pave the way for the collectibles industry while working aboard the Salyut and Mir space stations. The outposts contained a small "post office" where flown items could be stamped with an ink cancellation to certify that a certain artifact did fly in space. Shuttle astronauts are allowed to fit souvenirs into a grab bag called a personal preference kit. NASA tightly regulates what crew members can take with them and forbids them to sell items for profit.[4] Preservationists are worried that a determined lunar visitor could take capitalism a bit too far by venturing to the *Apollo 15* landing site to poach the lunar rover for profit. "Fifty years down the road, I don't want this stuff showing up on eBay," says O'Leary. The anguish among archaeologists goes beyond seeing Apollo artifacts on the auction block. The study of historic sites means examining artifacts precisely where they were left, or dumped in the case of the *Apollo 11* crew. Supporters of this cause are wading into a gray area of the law, for under international treaty no one owns the moon, so it is uncertain what authority the U.S. government has over the items left by American astronauts.[5]

Even if NASA returns to the lunar surface, that may not guarantee the security of the older Apollo landing sites. Early plans for NASA's first post Space Shuttle era Moon base means targeting the polar regions, which are considered to be cost effective, safe, and scientifically interesting destinations. Shackleton Crater at the South Pole is one proposed spot where long-term Astronaut expeditions could possibly harvest water ice for use during the visit.[6] If the agency follows through on this concept, visiting the old Apollo landing sites might be challenging, since they're clustered around the equator and not at the poles. That could make it difficult for 21st century NASA astronauts to travel great distances from Shackleton crater to put up "no trespassing signs."

New Space Program, Old Politics

When U.S. astronauts returned from joint missions aboard the Russian space station Mir, they would invariably be asked how the older Russian station would stack up against the new International Space Station. The response was invariably that everyone should wait a decade or so and see how the new station performed over the long haul. NASA has a lengthy path to build a new spacecraft that can make it all the way to Mars and back, and success is far from assured. The guidelines for the new moon-and-Mars initiative call for an ongoing mission of exploration, not a specific destination that signals the end of the program, like Apollo. This presupposes an ongoing flow of congressional funding, which in the past has been incremental. This could be a problem for NASA during its moon and Mars trips, as it was during the era of the shuttle and the space station. Dollars for the space program are doled out in fits and starts, as they are for most federal programs. That occasions furious debate among lawmakers seeking money for competing projects, many of which may be considered more valuable than spaceflight.

NASA originally envisioned the shuttle and space station as a matched set of projects, dependent on each other. Whether it was hesitation on the part of lawmakers or simply a distaste for another Apollo-sized budgetary boondoggle, the shuttle came first and the station much later. One wonders how the much maligned orbiters would have fared in the world of public opinion had they possessed a useful function by "shuttling" to and from a space station from the start.

As NASA moves into Project Constellation, the worst-case scenario would be for Capitol Hill to provide just enough funding to return to the moon and then halt the program. Images of twenty-first-century astronauts bouncing around the lunar surface in a rerun of Apollo would likely encourage the belief that the U.S. space program was accomplishing little more than what our grandparents did back in the 1960s.

One bit of wisdom in this ongoing debate comes from aerospace engineers. There is an old saying in their profession: "You can have something cheap, you can have it fast, or you can have it good. But you can only pick two." In other words, a spacecraft that's good in quality and is built fast won't be cheap. If it's good and cheap, it won't come about fast. And finally, if it's cheap and built fast, it won't be good.

How NASA ultimately fares during the new space program is probably a question to be answered by one of my journalistic descendants. That person will write the epitaph for Project Constellation, just as I've endeavored to do for the space shuttle. Only then will we know for sure.

Notes

Chapter 1. The Future

1. Carl Walz, interview by author, at NASA's 1st Space Exploration Conference, Orlando, February 1, 2005.

2. George E. Mueller, "The New Future for Manned Spaceflight Developments," *Astronautics and Aeronautics*, March 7, 1969, 24–32.

3. Pallab Ghosh, "Making Space Vision a Reality," BBC, January 19, 2006.

4. Mueller, "New Future."

5. Robert C. Seamans Jr., interview by author, June 2005.

6. Seamans, interview.

7. Planetary Society, "Extending the Human Presence in the Solar System," July 2004.

8. Ghosh, "Making Space Vision a Reality."

9. President John F. Kennedy, address at Rice University, September 12, 1962. Original text available at Woodson Research Center, Fondren Library, Rice University.

10. Isom "Ike" Rigell (NASA engineer), interview by author, 1999.

11. President George W. Bush, address to unveil "Vision for Space Exploration Program," NASA Headquarters, Washington, D.C., January 14, 2004.

12. Bob Keeling, "History Happened Here," *Orlando Magazine*, June 2004, 62–64.

13. NASA's 1st Space Exploration Conference, Orlando, January 30–February 1, 2005.

14. Ibid.

15. Press tour of Kennedy Space Center following *Challenger* disaster, March 1986.

16. See note 1.

17. Eileen Collins, interview by author, October 2003.

18. NASA briefing on lunar architecture, September 19, 2005.

19. Ibid.

20. Wayne R. Matson, ed., *Cosmonautics: A Colorful History*, 46–47.

21. Ibid.

22. Lockheed Martin press event on CEV, Florida Spaceport Authority, February 22, 2006.

23. Ibid.

24. "NASA Names the New Spacecraft Orion," BBC, August 23, 2006.

25. NASA Constellation Exploration update, Kennedy Space Center, August 2006.

26. Ibid.

27. Lee Morin, interview by author, September 15, 2006.

28. NASA, *NSTS 1988 News Reference Manual* (hereafter cited as *Manual*), <science.ksc.nasa.gov/shuttle/technology/sts-newsref/>, 73–75.

29. Morin, interview.

30. *NASA Space Shuttle News Reference*, 1998, 86–87.

31. Morin, interview.

32. Ibid.

33. Ibid.

34. John Glenn, interview by author, 1995.

35. Ibid.

36. John Gibbons, interview by author, June 2005.

37. Jay Barbree, interview by author, 1994.

38. Michael Griffin, NASA administrator, testifying on May 18, 2005, before the U.S. Senate Committee on Commerce, Science, and Transportation, Subcommittee on Science and Space, <http://legislative.nasa.gov/hearings/2005%20hearings/5-18-05%20Griffin.html>.

38. Charles "Pete" Conrad Jr., interview by author, 1994.

39. Gibbons, interview.

Chapter 2. New Spaceships, New Astronauts

1. STS-121 press briefing, Kennedy Space Center, February 2006.

2. Justine Hankins, "Lost in Space," *Guardian*, March 20, 2004.

3. Rigell, interview.

4. Zoya Zarubina, interview by author, 1994.

5. Ibid.

6. Dr. Nick Kanas, interview by author, June 2005.

7. Robert Crippen, interview by author, October 2006.

8. Gary Federici, "From the Sea to the Stars: A History of Navy Space and Space-Related Activities," June 1997, <www.history.navy.mil/books/space/index.htm>, chap. 2.

9. NASA, "Astronaut Selection and Training," NP-1997-07-006JSC, July 1997, <www.jsc.nasa.gov/pao/factsheets/>, p. 2.

10. Alan Bean, interview by author, 2002.

11. Carl Walz, interview by author, February 2005.

12. Ibid.

13. Marc Garneau, interview by author, November 1998.

14. Kanas, interview.

15. Victoria Coverstone, interview by author, June 2005.

Chapter 3. 1981: The Path to STS-1

1. Conrad, interview.

2. Seamans, interview.

3. Dave Typinski, "How Wings Work," Aerospace Education presentation, Gainesville Composite Squadron, Civil Air Patrol, July 21, 2005.

4. Seamans, interview.

5. Ibid.

6. Ibid.

7. Ibid.

8. Ibid.

9. Sam Beddingfield, interview by author, June 2005.

10. Crippen, interview.

11. John Young, interview by author, October 2006.

12. Terminal Countdown Demonstration Test, STS-93, June 1998.

13. Scott Horowitz, interview by author, October 2001.

14. Young, interview.

15. *Manual*, 73–77.

16. Dan Brandenstein, interview by author, June 2005.

17. Ibid.

18. *Manual*, 14.

19. *Manual*, 265–67.

20. Young, interview.

21. Ibid.

Chapter 4. 1982–1986: The Curse of Being Routine

1. Presidential Commission on the Space Shuttle *Challenger* Accident, *Report to the President*, 5 vols. (Washington, D.C., 1986) (hereafter cited as *Rogers Commission Report*), vol. 1, *Executive Summary*, chap. 6.

2. Beddingfield, interview.

3. Ibid.

4. *Manual*, 71–78.

5. Paul Bilke, "X-15 Research Results," NASA Flight Research Center, 1964.

6. Ibid.

7. Ibid.

8. *Rogers Commission Report*, vol. 1, chap. 6.

9. Brandenstein, interview.

10. *Gemini 5* crew patch, NASA photo S66-59530.

11. Frank Kuznik, "Personal Effects," *Air & Space/Smithsonian*, December 1994.

12. Ibid.

13. "The Crawler Transporters of Launch Complex 39," American Society of Professional Engineers, 1977.

14. Young, interview.

15. Brandenstein, interview.

Chapter 5. 1986: "Don't Call Unless It Blows Up"

1. Jenny Eschen Carter, interview by author, December 2004.

2. NASA press kit, STS-51L, January 1986.

3. Jenny Eschen, Robby Samson, and Jenny Golden, interview by author, January 28, 1986.

4. STS-51L crewmembers during training in flightdeck simulation, NASA photo S85-46207.

5. Theresa Manning, interview by author, June 2005.

6. George Diller, interview by author, January 28, 1986.

7. Press conference with Vice President George Bush and U.S. senators Jake Garn and John Glenn, January 28, 1986.

Chapter 6. The Investigation

1. Richard Cook, "The Rogers Commission Failed: Questions It Never Asked, Answers It Never Listened To," *Washington Monthly*, November 1986, 1.

2. *Rogers Commission Report*, preface, 1.

3. Press tour of *Challenger* wreckage, Kennedy Space Center public affairs, March 1986.

4. Terry Armentrout, interview by author, March 1986.

5. Ibid.

6. Ibid.

7. *Rogers Commission Report*, vol. 1, chap. 6, "An Accident Rooted in History," 120.

8. Ibid., "Early Tests," 122.

9. Ibid.

10. Ibid., "Leak Check and Putty," 133.

11. Ibid.

12. Ibid., "Criticality Classification and Changes," 121.

13. Ibid., "STS-51C and Cold Weather," 135.

14. Ibid.

15. *Rogers Commission Report*, vol. 1, chap. 3, "The Accident," 19.

16. Ibid.

17. *Rogers Commission Report*, vol. 1, chap. 6, "Findings," 148.

18. Ibid., "Flaws in the Decision Making Process," testimony of Roger Boisjoly, 89.

19. Ibid., testimony of Lawrence Mulloy, 100.

20. Ibid.

21. Ibid.

22. Cook, "The Rogers Commission Failed."

23. Joseph Kerwin to Richard Truly, July 28, 1986. <http://www.hq.nasa.gov/office/pao/History/kerwin.html>.

24. Ibid.

25. E. J. Tomei, "The Air Force Space Shuttle Program: A Brief History," *Crosslink*, Winter 2003, <http://www.aero.org/publications/crosslink/winter2003/05.htm>.

26. Roger Guillemette, "The Curse of the 'Slick 6': Fact or Fiction?" *Florida Today*, May 10, 1999.

27. Marcia S. Smith, "Space Launch Vehicles: Government Activities, Commercial Competition, and Satellite Exports," 2003, <usinfo.state.gov/usa/infousa/tech/space/launch.pdf>, p. 5.

Chapter 7. 1988–1992: Back on the Horse

1. John Lounge, interview by author, February 2005.

2. Dick Covey, interview by author, 2003.

3. Lounge, interview.

4. Mission audio STS-26, NASA PAO Johnson Space Center, 1988.

5. Brandenstein, interview.

6. Mark Carreau, "10 Years after Challenger, NASA Says Shuttle Safety Never Better," *Houston Chronicle*, January 19, 1996.

7. Jim Banke, "NASA Develops New Plans for Shuttle Crew Safety," *Space News*, July 7, 2003.

8. Brandenstein, interview.

9. William Broad, "Newest Shuttle, Endeavour, Begins First Mission," *New York Times*, May 8, 1992.

10. European Space Agency, *The Ulysses Data Book* (Paris, 1990), 1.

11. David Borowsky, "Ulysses," 1992, <www.tsgc.utexas.edu/archive/characterizations/ulysses2.html>.

12. Bruce Melnick, interview by author, June 2005.

13. Ibid.

14. NASA press kit, STS-41.

15. Ibid.

16. NASA press kit, STS-49.

17. Boeing, "Intelsat VI F-3: Major Points in the Mission," <www.boeing.com/defense-space/space/bss/factsheets/376/intelsat_vi_f3/intelsat_vi_f3.html>.

18. *Manual.*

Chapter 8. 1993: The Road to Mir

1. Gibbons, interview.

2. Ibid.

3. Roald Sagdeev, interview by author, October 2006.

4. David R. Williams, "Apollo 18 through 20: The Cancelled Missions," NASA Goddard Space Center, August 2000, <http://nssdc.gsfc.nasa.gov/planetary/lunar/apollo_18_20.html>.

5. Glenn Garelik, "Our Boys Are Dead," *Discover*, April 1986.

6. Robert Zimmerman, "Docking in Space: Years of Danger and Heroics That Made It Possible," *Invention and Technology*, Fall 2001.

7. Sagdeev, interview.

8. Ibid.

9. Michael L. Cheatham, *Advanced Trauma Life Support for the Injured Astronaut*, 3rd ed. Orlando: Orlando Regional Medical Center, December 2003.

10. Sagdeev, interview.

11. Matson, *Cosmonautics*, 51–76.

12. Ibid.

13. Janne Nolan, "US-Russian Engagement in Space: The View from the United States," The Eisenhower Institute, August 20, 2002, <www.eisenhowerinstitute. org/programs/globalpartnerships/fos/newfrontier/US-Russian%20Engagement% 20in%20Space%20-%20The%20View%20from%20the%20United%20States%20 %20Janne%20Nolan.pdf>.

14. Sagdeev, interview.

15. Norman Thagard, interview by author, October 2006.

16. Gibbons, interview.

17. NASA TV coverage of STS-63, 1994.

18. Vance Brand, interview by author, 1994.

19. Robert "Hoot" Gibson, interview by author, 1995.

20. Thagard, interview.

21. Matson, *Cosmonautics*, 51–76.

22. Thagard, interview.

23. Gibson, interview.

24. Gen. Thomas P. Stafford Task Group, "First Joint Report on Shuttle/Mir Rendezvous and Docking Missions," *First Joint Report on Shuttle/Mir Docking Flights*, June 27, 1996, 23.

25. NASA press kit, STS-71, 1995.

26. Charles Precourt, interview by author, 1997.

27. NASA press kit, STS-74, 1995.

28. NASA TV coverage of STS-74, 1995.

29. Ibid.

30. NASA press kit, STS-81.

31. Jerry Linenger, interview by author, October 2006.

32. Ibid.

33. Ibid.

34. Ibid.

Chapter 9. 1998: ISS, the Shotgun Marriage

1. Sagdeev, interview.

2. Congressman Tim Roemer, interview by author, 2000.

3. Dana Rohrbacher, chairman, U.S. House Subcommittee on Space and Aeronautics, opening statement to "International Space Station: Status and Cost Overruns," November 5, 1997, <http://commdocs.house.gov/committees/science/hsy309160.000/hsy309160_0.htm>.

4. Douglas Holtz-Eakin, "Budgetary Analysis of NASA's New Vision for Space Exploration," Congressional Budget Office, September 2004.

5. Hal J. Strumpf, Vahe Avanessian, and Reza Ghafourian, "Design Analysis and Life Prediction for the Brayton Engine Receiver for the Space Station Freedom Solar Dynamic Option," in *Proceedings of the 26th Intersociety Energy Conversion Engineering Conference* (La Grange Park, Ill.: American Nuclear Society, 1991), 1:241–47.

6. Gibbons, interview.

7. Marcia S. Smith, "Space Station Freedom," issue brief for Congress, February 3, 2003.

8. Glenn Snyder, interview by author, November 1998.

9. Virginia Barnes, interview by author, November 1998.

10. Ibid.

11. Snyder, interview.

12. NASA press kit, STS-88, 1998.

13. Snyder, interview.

14. Stealth Dog Crew 3 patch design, <www.spacepatches.nl/sts_mis/STS88a.html>.

15. Bob Cabana, interview by author, 1999.

16. U.S. House, Committee on Science, debate over HR 106-145, National Aeronautics and Space Administration Authorization Act of 1999, May 18, 1999, Congressional Record.

17. Ibid.

18. Jacob J. Lew, director, Office of Management and Budget, testifying before the U.S. House Committee on Science, August 6, 1998, Congressional Record.

Chapter 10. *Columbia*

1. "Joint Press Conference—Bill Clinton, Shimon Peres," Israeli Ministry of Foreign Affairs, December 11, 1995, <www.mfa.gov.il/MFA/Archive/Speeches/JOINT%20PRESS%20CONFERENCE%20CLINTON-PERES%20-%2011-Dec-95>.

2. Ibid.

3. Nathan Guttman, "Remains of Ilan Ramon Identified, to Be Buried in Israel," *Haaretz*, June 2, 2003.

4. Craig Covault, "Israeli Astronaut, Diverse Science Launched Amid Tight Security," *Aviation Week & Space Technology*, January 20, 2003.

5. RMIT University (Melbourne, Australia) press release, "Spiders in Space," January 21, 2003.

6. Associated Press, "Columbia's Problems Began on Left Wing," *Houston Chronicle*, February 1, 2003.

7. NASA, Columbia Accident Investigation Board, *Report* (Washington, D.C., 2003) (hereafter cited as *CAIB Report*), vol. 1:77.

8. William Harwood, "Atlantis Returns to Earth," CBS, October 18, 2002.

9. Johnson Space Center PAO, "STS-107 Audio Closed Communication Loop," February 1, 2003.

10. Gwen Thompkins, interview by author, June 2005.

11. D'Arcy Bacon, interview by author, June 2005.

12. Steve Lindsey, interview by author, October 2005.

13. Sheryl Chaffe Marshall, interview by author, January 2006.

Chapter 11. The Aftermath

1. Stefano Coledan, "Columbia: The Final Report," *Popular Mechanics*, August 26, 2003.

2. *CAIB Report*, testimony of Harry McDonald, March 6, 2003, 6:32.

3. Ibid.

4. Ibid.

5. Ibid.

6. *CAIB Report*, testimony of Aloysius Casey, March 25, 2003, 6:129.

7. Arthur Baker, interview by author, November 2003.

8. Ibid.

9. Ibid.

10. Ibid.

11. Ibid.

12. Todd Halvorson, "Shuttle Columbia's Flight Data Recorder Found," *Florida Today*, March 19, 2003.

13. Press tour of *Columbia* wreckage hangar, Kennedy Space Center, July 2003.

14. *CAIB Report*, "Assessment of Potential Material Candidates for the 'Flight Day Two' Radar Object Observed during the NASA Mission STS-107," vol. 3, app. E3.

15. See note 13.

16. Press conference upon release of *CAIB Report*, NASA Headquarters, Washington, D.C., August 26, 2003.

17. Ibid.

18. John Clark, interview by author, October 2003.

19. Ibid.

Chapter 12. Back on the Horse, Again

1. Margaret Conklin, interview by author, January 2005.

2. Eileen Collins bio, Johnson Space Center, August 2005.

3. Associated Press, "Too Short Astronaut Misses Mir," *Augusta Chronicle*, September 15, 1997.

4. STS-114 crew fit-check press briefing.

5. Chris Kridler, "Shuttle Shuffle: Discovery Gets the Call," *Florida Today*, July 6, 2005.

6. David S. F. Portree, "Mir Hardware Heritage," NASA RP 1357, Johnson Space Center, 1995.

7. STS-100 Terminal Countdown Demonstration Test press briefing, Kennedy Space Center, Florida, February 2001.

8. Soichi Noguchi, interview by author, October 2003.

9. Ibid.

10. Paul Rincon, "NASA Investigates Shuttle Failure," BBC, July 14, 2005.

11. Kelly Young, "Falling Foam Grounds the Shuttle," *New Scientist*, July 28, 2005.

12. Pam Easton, "Discovery Astronaut Neatly Tugs Both Pieces of Fabric from Shuttle's Belly," Associated Press, *Deseret News*, August 4, 2005.

13. Marcia Dunn, "Shuttle Astronauts Depart ISS, Prepare for Anxious Trip Home," Associated Press, *The Hawkeye*, August 6, 2005.

14. Lindsey, interview.

15. Mission management team briefing on PAL ramp foam loss on STS-114, Johnson Space Center, Houston, Texas, July 27, 2005.

Chapter 13. The Moon, Maybe, but Mars?

1. Jim Gerard, "A Field Guide to American Spacecraft," Oklahoma State University, <http://aesp.nasa.okstate.edu/fieldguide>.

2. Charles Duke, interview by author, 1994.

3. Patricia Downey, interview by author, February 2005.

4. International Space Station logistics press briefing, Johnson Space Center, November 2003.

5. Steven W. Squyres, interview by author, June 2005.

6. Ibid.

7. Kanas, interview.

Epilogue

1. Conrad, interview.

2. James Knight, "China Near First Human Spaceflight," *New Scientist*, September 16, 2003.

3. Beth O'Leary, interview by author, June 2005.

4. Kuznik, "Personal Effects."

5. O'Leary, interview.

6. NASA Global Exploration Strategy and Lunar Architecture Briefing, December 2006, NASA Headquarters, Washington, D.C.

Suggested Reading

Aldrin, Buzz, and Malcolm McConnell. *Men From Earth*. New York: Bantam, 1989.

Cooper, Henry S. F., Jr. *The Evening Star: Venus Observed*. New York: Farrar, Straus & Giroux, 1993.

Harland, David M. *The Story of the Space Shuttle*. New York: Springer, 2004.

Joels, Kerry Mark, and Gregory P. Kennedy. *Space Shuttle Operator's Manual*. Rev. ed. New York: Ballantine, 1988.

Johnson-Freese, Joan. *The Chinese Space Program: A Mystery within a Maze*. Malabar, Fla.: Krieger, 1998.

Kerrod, Robin. *NASA: Visions of Space*. Philadelphia: Courage Books, 1990.

Lebedev, Valentin. *Diary of a Cosmonaut: 221 Days in Space*. Translated by Luba Diangar. Edited by Daniel Packett and C. W. Harrison. College Station, Tex.: PhytoResource Research, 1988.

Linenger, Jerry M. *Off the Planet: Surviving Five Perilous Months Aboard the Space Station Mir*. New York: McGraw-Hill, 2000.

Lovell, Jim, and Jeffrey Kluger. *Lost Moon: The Perilous Voyage of Apollo 13*. Boston: Houghton Mifflin, 1994.

Matson, Wayne R., ed. *Cosmonautics: A Colorful History*. Rockville, Md.: Cosmos Books, 1994.

Murray, Charles, and Catherine Bly Cox. *Apollo: The Race to the Moon*. New York: Simon and Schuster, 1989.

Reichhardt, Tony, ed. *Space Shuttle: The First 20 Years*. New York: DK Publishing, 2002.

Shelton, William R. *Man's Conquest of Space*. 4th ed. Washington, D.C.: National Geographic Society, 1975.

Thompson, Milton O. *At the Edge of Space: The X-15 Flight Program*. Washington, D.C.: Smithsonian Institution Press, 1992.

Index

Award-winning journalist Pat Duggins is a contributing correspondent to National Public Radio in Washington, D.C., where he reports on Florida and the U.S. space program. He is also the senior news analyst for Orlando's NPR station, 90.7 WMFE-FM, and adjunct professor of journalism at the University of Central Florida's Nicholson School of Communication.

Related-interest titles from University Press of Florida

"Before This Decade is Out . . ." Personal Reflections on the Apollo Program
Edited by Glen E. Swanson

Big Dish: Building America's Deep Space Connection to the Planets
Douglas J. Mudgway

Florida's Space Coast: The Impact of NASA on the Sunshine State
William Barnaby Faherty

Gateway to the Moon: Building the Kennedy Space Center Launch Complex
Charles D. Benson and William Barnaby Faherty

A History of the Kennedy Space Center
Kenneth Lipartito and Orville R. Butler

Moon Launch!: A History of the Saturn-Apollo Launch Operations
Charles D. Benson and William Barnaby Faherty

The Soviet Space Race with Apollo
Asif A. Siddiqi

Stages to Saturn: A Technological History of the Apollo/Saturn Launch Vehicles
Roger E. Bilstein

For more information on these and other books,
    visit our Web site at www.upf.com.